AF802676

The Management of Construction Safety and Health

Edited by
RICHARD J. COBLE, THEO C. HAUPT
& JIMMIE HINZE

CRC Press
Taylor & Francis Group
Boca Raton London New York

CRC Press is an imprint of the
Taylor & Francis Group, an **informa** business
A BALKEMA BOOK

Published by:
CRC Press/Balkema
P.O. Box 447, 2300 AK Leiden, The Netherlands
e-mail: Pub.NL@taylorandfrancis.com
www.crcpress.com – www.taylorandfrancis.com

© 2000 by Taylor & Francis Group, LLC
CRC Press/Balkema is an imprint of the Taylor & Francis Group, an informa business

No claim to original U.S. Government works

ISBN 13: 978-90-5809-328-8 (hbk)

This book contains information obtained from authentic and highly regarded sources. Reasonable efforts have been made to publish reliable data and information, but the author and publisher cannot assume responsibility for the validity of all materials or the consequences of their use. The authors and publishers have attempted to trace the copyright holders of all material reproduced in this publication and apologize to copyright holders if permission to publish in this form has not been obtained. If any copyright material has not been acknowledged please write and let us know so we may rectify in any future reprint.

Except as permitted under U.S. Copyright Law, no part of this book may be reprinted, reproduced, transmitted, or utilized in any form by any electronic, mechanical, or other means, now known or hereafter invented, including photocopying, microfilming, and recording, or in any information storage or retrieval system, without written permission from the publishers.

For permission to photocopy or use material electronically from this work, please access www. copyright.com (http://www.copyright.com/) or contact the Copyright Clearance Center, Inc. (CCC), 222 Rosewood Drive, Danvers, MA 01923, 978-750-8400. CCC is a not-for-profit organization that provides licenses and registration for a variety of users. For organizations that have been granted a photocopy license by the CCC, a separate system of payment has been arranged.

Trademark Notice: Product or corporate names may be trademarks or registered trademarks, and are used only for identification and explanation without intent to infringe.

Visit the Taylor & Francis Web site at
http://www.taylorandfrancis.com

and the CRC Press Web site at
http://www.crcpress.com

Contents

Preface

The construction industry still clings to its infamous position as the industrial sector responsible for more occupational accidents, injuries and fatalities than any other. Consequently, the improvement of safety and health in construction is still a necessary goal for all participants in the construction process from project inception with the desire of a client to erect a building, through to the final demise of the facility through demolition at the end of its lifespan. There is a dynamic international drive to redistribute the responsibility for safety in construction away from contractors, who traditionally were solely responsible for safety on construction sites, to include building owners, designers, engineers, suppliers, manufacturers, workers and end users. This is a positive initiative, which deserves the support of the entire industry since safety is everyone's business!

This text is a contribution towards the effort to improve safety performance in construction. It has been produced from a selection of the papers presented at the University of Florida in the United States during the Rinker Eminent Scholars Lecture Series on Construction Safety and Health from August through December 1998. Each of the scholars has built up an international reputation for their work in specific areas of construction safety. The chapters in this book are produced specifically for all participants in the construction process, students and others who have more than a passing interest in producing an industry which is safe for all. The many techniques and approaches presented in the book are practical and can be implemented with ease by any construction participant.

In his chapter on behaviour measurement for continuous improvement, Roy Duff from the United Kingdom, suggests that the measurements can be used to focus attention on those areas of operation requiring improvement, as baseline measures for setting improvement targets, as measures of the improvements achieved, and as objective benchmarks of performance being achieved by the best organisations. He discusses experiences in the development of measures of safety on construction sites, with particular reference to concrete construction and masonry. Improvements over 15 construction sites averaged 23%.

Luís Alves Dias from Portugal postulates that the enactment of a significant number of laws and regulations driven by the World Health Organisation (WHO), International Labour Organisation (ILO) and International Organisation for Standardisation (ISO), reflects the world concern for the reduction of the high number of occupational accidents and illnesses. In his chapter on the management of safety, health and the environment in construction, Dias argues that the practical and successful implementation of the requirements of these laws and regulations requires management approaches which vary according to the specificity of the industry for which they must be applied. He recommends the integration of environmental, quality and safety management systems if the lives of workers, and the environment are to be preserved.

In his chapter focusing on the Council Directive 92/57/EEC, Joachim Berger from Germany, stresses that in none of the 12 member states of the European Community is it possible to sit back and claim that safety and health protection on construction sites has reached acceptable levels. He discusses the implications of the Directive for encouraging improvements in the working environments of construction workers.

A structural reliability analysis method is proposed by Xila Liu from the People's Republic of China. In his chapter on the analysis and control of human errors during structural construction, he suggests that human errors occurring during the planning, design and construction of structures are the main causes of structural failures. He produces a simulation model by means of which it is possible to analyse the influence of these errors and to control the errors more efficiently.

According to Michael Baxley from the United States, health risk management in construction poses unique challenges due to the nature of the industry. He argues for a strong focus on prevention, safety and early recognition as the keys to optimal health risk management on an integrated basis in which labour and management each play an important role.

Pam Susi from the United States, in her chapter on the principles of exposure assessment, presents a strong argument that the degree to which workers are at risk of disease and premature death is dependent on their exposure to hazardous agents. A 'systems' approach to controlling hazards is needed. Exposure assessment and control efforts must be part of more comprehensive safety and health programmes to be effective.

In his chapter in which he takes an overview of health hazards in construction, Robert Herrick from the United States, suggests that health hazards can be recognised, evaluated, and controlled, provided that the unique nature of construction is recognised. He presents three approaches to health hazard control. These are the identification of hazards by construction type, linking of hazards to each of the major trades and crafts prevalent in construction and the identification of a set of exposures. He argues that such a three-dimensional approach will result in a comprehensive assessment of construction-related health hazards.

Ron Sharpe from Australia, indicates that information technology is helping to improve health and safety. In his chapter on the use of IT to help regulate public health and safety in construction, he suggests that IT presents opportunities for improving safety both for workers and users. Unsafe environments, for example, may be avoided through the use of automation and robotics.

In similar vein, Theo Haupt from South Africa, John Alexander and Richard Coble from the United States, in their chapter on automated technology in construction safety, present several of the latest developments in automation and robotics which have the potential to improve levels of safety and reduce exposure to hazards on construction sites.

Pierce Jones from the United States, builds on the ability of computers to manipulate, store and retrieve vast amounts of data, and presents a hazard communication system using interactive computer based training techniques. The training modules are designed to be highly interactive and generally operate like games.

It is hoped that this book will help in the all-out war against accidents, injuries and fatalities in the construction industry and produce a safer industry. The words of Barbara De

Woody of Universal Studios, Florida that 'safety is everyone's business' ring truer today than ever before. They need to ring louder still!

The editors acknowledge the contributions of every scholar in the fight to improve safety and health in construction. Without the efforts of Roy Duff, Luís Alves Dias, Pam Susi, Michael Baxley, Ron Sharpe, Xila Liu, Joachim Berger, Bob Herrick, Pierce Jones, John Alexander, Theo Haupt and Richard Coble, this book would not have been produced. Thank you also to the thousands of warriers who wage war daily to keep construction safe.

Richard Coble
Theo Haupt
Jimmie Hinze

Gainesville, Florida, USA

June 2000

CHAPTER 1

Behaviour measurement for continuous improvement in construction safety and quality

A. ROY DUFF
Department of Building Engineering, UMIST, UK

ABSTRACT: Continuous improvement is now a widely accepted part of organisational and industrial initiatives, such as Total Quality Management, Partnering and Benchmarking. Its objective is to focus attention on those aspects of the operation of the organisation that will add value to the customer and to direct the activities of the organisation such that performance is being continuously monitored and improved. The fundamental requirement of this approach is a satisfactory means of measuring performance. If this can be achieved, then the measurements which are taken can be used in a variety of ways: as an organisational focus on those areas of operation requiring improvement; as baseline measures for setting improvement targets; as measures of the improvements achieved; and as objective benchmarks of performance being achieved by the best organisations.

Measurement of performance, if it is to be effective in an industrial context, must have a variety of features. For example: it must be a valid representation of the parameter of performance which requires improving; it must be reliable, in that the same level of performance will produce the same score, if repeated; it must be acceptable to the people whose performance is being monitored. The chapter discusses these and other issues in performance measurement.

The chapter also describes experiences in the development of such measures of safety and quality on construction sites. Safety measurement included scaffolding, other access to heights, site housekeeping, wearing of personnel protective equipment and operation of construction plant. These were used to determine baseline performance and to set targets for improvement. The continuous improvement programme was based on the techniques of goal-setting and feedback from organisational psychology. Improvement averaged 23% over 15 construction sites. Special measures were also developed for the correct selection, wearing and maintenance of specialist respiratory equipment, and for construction operations adjacent to live motorways.

Measurement of quality provides many difficulties, not least of which are those arising from the very subjective nature of many of the parameters of quality, such as quality of finish or aesthetic quality. The chapter describes experience in the development of two different measures of construction quality, in concrete construction and masonry. The objectives of the measurement process are shown to produce two significantly different approaches and the reasons for the difference are discussed.

1 INTRODUCTION

There is much talk in the construction industry about the need to adopt the philosophy of continuous improvement and most construction companies are now recognising this need. However, it is vital also to recognise that any performance improvement programme requires careful design, implementation and commitment, if it is to be successful. There are a number of important principles that have to be embraced, if the desired improvement goals are to be achieved. This chapter considers the importance of performance measurement in efforts to improve performance in construction, the principles that must be adopted to achieve effective measurement of performance, and examples of some recent UMIST research and development in the performance measurement of construction safety and quality.

The globalisation of all industries is raising the pressure on construction to respond to increasing international competition and client demands for higher and higher productivity and quality, together with lower costs and project durations. Construction companies are no longer protected by the geographic isolation of their activities. The larger clients, with operations around the world, are able to make direct comparisons between the service they receive from the construction industries of several countries. Preliminary findings of a recent client survey in the UK (New Civil Engineer, 3 December 1998, p7) suggest that: only 39% of projects are delivered on or ahead of time; average time overrun was 14.8%; overall project satisfaction was 7.7 out of 10. Construction companies, or even national industries, which do not perform are now vulnerable to the threat of being replaced, or taken over, by those that do.

This internationalisation of competition has exposed the construction industries of the world to the sort of pressure that has been commonplace in other engineering industries for decades. All manufacturers must continually offer more technical sophistication, more added value, better service and lower cost to the customer, just to remain in business. Continuous improvement is no longer a theoretical ideal, it is a commercial necessity. Construction companies, like all other trading organisations, must improve their performance, year on year, just to survive.

2 THE NEED FOR PERFORMANCE MEASUREMENT

There have been several movements, successful in other engineering manufacturing industries, which are seen to have the potential to improve the performance of construction. All of them embrace, in one way or another, the goal of performance improvement. Total Quality Management has continuous improvement as one of its fundamental concepts, the internal market being the other. Partnering involves developing commercial and operational relationships between client and contractor, or within the supply chain, with the prime objective of progressively improving the performance of the whole project delivery process. The essence of the relationship is a voluntary, non-contractual agreement to work together, in collaboration, to achieve agreed targets for improvement and their associated benefits. The benefits are shared by all parties in the partnership. Benchmarking has the objective of identifying best practice and mapping the processes that constitute best practice, all with a view to transferring this practice to under-performing organisations. In each case, the tools enabling organisations to achieve the objective of continuously improving performance, and hence competitiveness, must include methods of performance assessment.

In order to be meaningful, the goal of continuous improvement must be objective, progressive, quantified and possible to monitor. This means measurement; subjective assessment is not adequate. TQM requires measures of performance to provide focus for improvement effort and to enable improvement to be monitored. Partnering requires accepted goals, at which the partners can all aim, to give the partnership both unity of purpose and, in the case of efficiency or profit related goals, a basis for distributing the benefits. Benchmarking assumes that best practice can be identified. Without measures of performance, this identification will be subjective and subject to individual perception, and even fashion, rather than driven by the search for real efficiency and hard commercial reality.

Many writers have embraced the idea of measurement of performance as a means to improve the management of organisations, none more succinctly than Peter Drucker, who wrote simply – 'What gets measured gets managed.' The whole philosophy of identification of the important parameters of organisational performance and success has more recently been encapsulated in the phrase 'Key Performance Indicator' and most well managed organisations now identify and communicate their KPIs as a fundamental part of their management process. Unfortunately, some KPIs are more meaningful to the CEO, or Board of Directors, than they are to the people whose job it is to produce the results. Management by Objectives, a similar approached enthusiastically in the 1960's was fine, in theory, but its implementation failed too often because the objectives identified were neither specific enough, nor sufficiently proximate, in time or scope, to the people whose motivation was required for success.

Performance measurement has a variety of uses and great value. First, and possibly most important, it identifies and signals very clearly those aspects of the operation of the organisation that the management considers to be of high priority. It provides focus in the management process. It provides baseline standards against which any change can be objectively assessed. It can form the basis for objective appraisal of individual or group performance. It provides the focus for improvement goals, and feedback of progress towards those goals. Feedback, in itself, can provide significant reward, or 'punishment' to personnel who have a stake in the performance being measured. Its effect as the source of positive reinforcement of desired behaviours can be considerable. In itself, it can strengthen the reward of intrinsic job satisfaction and, when supported by externally given psychological rewards, such as supervisor praise or recognition by colleagues, it becomes a very powerful motivator.

On a broader front, performance measurement is the only objective basis for intra- or inter-organisational comparison and is, therefore, the only sound basis for the identification of best practice and benchmarking. It also has the power to encourage informal competition between groups and organisations. The UK Government is promoting a national Construction Best Practice Programme, which includes £6M of support for the identification of 100 plus demonstration projects which will be subject to performance measurement in the perceived key areas, as industry benchmarks.

3 CRITERIA FOR EFFECTIVE PERFORMANCE MEASUREMENT

Although the idea of performance measurement as the basis for performance improvement programmes has been around since the advent of scientific management, it is only comparatively recently that the power of specific performance goals has been scientifically investigated and the way they work properly understood. The criteria for their success are many and important, if the most effective use is to be made of goal-based improvement initiatives.

Operationally, goals affect performance in a number of ways. They focus attention on the aspects of performance that the management of the organisation perceives to be important. They mobilise, direct and sustain effort towards appropriate levels of performance, *providing they are accepted and there is commitment to their achievement.* Feedback of performance, sometimes called knowledge of results, is strongly related to the impact of goals in performance improvement. It forms the bases for the great majority of rewards, intrinsic or extrinsic, that emanate from the process. Intrinsic rewards, those that give pleasure and satisfaction from the job itself, are personal and feed the process of improving performance related behaviour by reinforcing the behaviour that results in reward and tending to extinguish the behaviour that does not. Extrinsic rewards can range from recognition of improved performance through praise, from supervisor or manager, to material reward in the form of a prize or performance-related pay. However, the performance measurement and feedback can, itself, seriously undermine the beneficial effect of this reward process if it fails to satisfy a number of criteria, necessary for acceptance of the measure by those whose performance is being evaluated and rewarded.

It may seem to be stating the obvious to say that measures of performance should be *quantifiable*. It is, however, beneficial to understand the reasons for this. Quantification of performance provides a clear and explicit scale of achievement and improvement, or deterioration. To describe a performance as adequate or good is subjective and ill-defined, compared with giving it a score of 50% or 70%. The score discriminates clearly and proportionally between two levels of performance.

Quantification of performance is most reliable and effective when it arises from actual measures, such as productivity measured by output in a fixed time, in which case the measure is seen as objective. In this case there is little room for subjective judgement and the accuracy of the result can be tested to support its acceptability. In some cases, however, this clear process of quantification by direct measurement cannot be achieved. Subjective judgement is unavoidable, for example, in assessing aspects of site housekeeping (tidiness) in measurement of the safety of a construction site, or in measurement of aspects of quality of finish in masonry work. In these cases, the approach should be to define the criteria for assessment as clearly, and in as much detail, as possible. Clarity of definition will allow experienced observers to arrive, systematically, at similar, if not identical, results that can be scored and hence converted to quantified measurements. Establishing as much detail as possible will allow the measure to be broken down into a number of parameters. Although these may each involve subjectivity in their assessment, the central limit theorem from mathematics, or the commonly understood notion of 'swings and roundabouts', suggest that the aggregate measure will be more accurate and reliable than the assessments of the individual parameters; and certainly much more reliable than a single overall assessment. In fact, it could be said that construction estimators and planning engineers only retain their jobs by virtue of this effect, which allows compensation of negative and positive errors.

Measures of performance must be *valid*, that is they must measure, and be seen to measure, what they are intended to measure. If construction site operatives do not believe that a tidy site contributes to site safety, then including a measure of site housekeeping (tidiness) in safety performance will lead to a perception that the measure is not a valid way of measuring their performance. This will, in turn, lead to a rejection of any goals or feedback associated with the measure. Improving performance will be seen to have little meaning or value, and the almost inevitable reduction in performance to be of little account. Validity has two aspects: real and face validity. If, for example, we can show that site housekeeping is directly related to the number of accidents arising from trips and falls, then it has real validity as part of a measure of safety performance. If, however, we can only imagine this connection, based on our own experience and evaluation of the likely outcome of an untidy site, then the measure has only face validity. It appears to measure what we want to measure. In many cases, fortunately, the logic of the situation is so clear that face validity is an adequate test of the validity of the measure. It is, after all, what we perceive to be true that governs our actions and, therefore, how we respond to any performance improvement initiative.

The measure, and the measurement process, must also be *reliable*. This implies that repeated use of the measure should yield the same result in the same circumstances. Significant changes in the measure of performance, in situations in which those being assessed can perceive no difference, will suggest errors in the measurement process and quickly lead to loss of confidence and a failure to accept the results. Equally, management must be confident in the reliability of performances measures if they are to be used as a basis for performance monitoring and management control.

Performance measures need to be *sensitive* but *stable*. Sensitivity and stability are related concepts in the design and operation of any measures. Sensitivity refers to the ability of the measure to respond to a real change in performance; stability to its ability to avoid large swings in the measure in situations of relatively small, or no, changes in actual performance. Suppose, for example, we assessed the measure of quality as, simply, good or poor and scored it: good = 10 points; and poor = 0 points. Then, a small change in quality at the threshold level, equivalent to the change from good to poor, would yield a large swing in the measure, whereas a relatively large change in quality at any other level would yield no change in the performance measure. In the first case, the measure of performance, at this level, would be too sensitive to a change in performance and tend to produce a very unstable result. In the second case, it would not be sensitive enough to give an adequate response, and hence reward, for a significant improvement, possibly after considerable effort on the part of the workforce. The negative impact on motivation and commitment to improvement goals, of either of these circumstances, can be anticipated. Any performance measure must be both stable enough, to avoid wild fluctuations, and sensitive enough, to respond to any significant change, over the whole range of expected performance.

Any measure of performance must be *timely*, to be effective as a motivator to higher performance. The prime objective of performance feedback is to increase the quality and frequency of desired, productive behaviour and to reduce the frequency of undesired, unproductive behaviour. This occurs through intrinsic or extrinsic systems of reward (or punishment). The impact of either is greatly reduced if it is received long after the behaviour that produced the performance and earned the reward. This requirement implies that the process of measurement and feedback should be quick. For example, a measure of quality that relies on immediate assessment of the output of the process will reward, giving positive reinforcement of high application of skill, and permit remedial measures if performance falls. If the measure relies on testing or inspection that can, or does, only take place at some later time, then reward for good performance has little effect on subsequent behaviour, and punishment is more likely to result in irritation, confrontation or withdrawal of co-operation, than in behaviour and performance improvement. The time lag in the management control process will also make it likely that sub-standard performance continues, unchecked, for an unacceptable length of time. Cost monitoring and control systems have often suffered from these deficiencies, due to the difficulty of maintaining up-to-date, reliable cost records.

Both the process of measurement and the measure itself must be *understandable* by the workforce, if they are to be accepted as a basis for performance assessment and for improvement goals to which commitment is to be obtained. The measurement process must be transparent and well within the comprehension of those being measured. A measurement process which requires complex equipment or calculation is likely to raise suspicion, particularly if the result does not appear to reflect the perception of quality, or committed effort, of the subjects. For a positive response, acceptance of the result is essential and requires trust in the measurement process.

Ideally, the measure of performance should be *generalisable*, over different locations (sites, organisations), if maximum benefit is to be obtained. A measure developed for a single location can be used as the basis for an improvement programme, but clearly has its drawbacks. It may be used to demonstrate improvement over time, but cannot be an absolute measure of performance, and hence cannot be used for inter-site comparison, as the basis of benchmarking or as a tool to monitor the success of the transfer of best practice. Without this ability, it will be impossible to benefit from the lessons learned on one site or project, in order to improve performance on the next. The development cost of such a one-off, customised performance measure will also impact negatively on the next requirement for effective performance feedback – its economic viability.

A performance measurement and improvement programme is an investment project and must be subject to the same *economic* justification as any other investment project. Clearly, it must be demonstrable that the economic benefits arising from any performance measurement programme are greater than the development, installation and operating costs. This implies that development costs that can be reduced by applying the results of external development programmes, or by spreading costs over many locations, long periods, or even over several organisations such as partnerships or like-minded 'clubs' are a great advantage. Whilst commercial advantage, of one organisation over another, will always be the essence of industrial competition, it is being increasingly recognised, particularly by clients, of which governments are the most influential, that a general improvement in performance is beneficial. One way of achieving this is by supporting the development of performance improvement initiatives in limited areas and using these as examples, to provide new benchmarks for the rest of the industry. This argument is being increasingly used to justify new research and development programmes such as the UK Construction Best Practice Programme.

An essential element of any system of management is the control of employee and, if possible, external participant behaviour. If continuous improvement is to be achieved, then people in the overall system of production must behave differently. Improvement in construction requires operatives, construction managers, designers, or even clients, to do things better. This can be achieved in a variety of ways but, in the absence of dictatorial control, people must be persuaded, motivated, to change their behaviour. Feedback, particularly when accompanied by goal-setting, is a behaviour change process. It therefore follows that it should be aimed, as closely as possible, at the *target behaviours*. Immediate reinforcement of desired behaviour by positive feedback, for example through praise or material reward, or 'punishment' of unwanted behaviour, will be far more effective than delayed assessment of, and managerial response to, the outcomes of the behaviour. This fact presents a problem for designers of performance measurement programmes. KPIs derived directly from commercial imperatives are not usually specified in terms that relate directly, and clearly, to the identification and reinforcement of optimum behaviour. A sales executive will clearly contribute most effectively to company profits by increasing sales but, unless it is clearly understood what improvement in behaviour is required to achieve this goal, a sales target is likely, on its own, to be ineffective as a motivator. Behaviour improvements, such as making more cold calls and promoting confidence in the product through obtaining more reference sells, may be the keys to improved performance and it is these behaviours that should be the focus of performance measures, and the basis for KPIs. In addition, to measure performance by outcomes will frequently cause a time lag in the feedback. This further diminishes the impact.

Measurement and feedback of performance is a fundamental part of any management control system and it is easy to believe that any performance feedback must be a good thing. However, there are important considerations. Feedback of data indicating poor performance will normally be seen as a criticism of the target behaviours. It is, after all, a corrective response to this implied criticism that is sought. It is important, therefore, that the performance being reported is *within the control*, or at least within the influence, of the people in receipt of the feedback. The whole process is related to the principles of delegation and one of the paramount principles of delegation is the exact match of authority and accountability. Performance feedback implies accountability and one should never be held accountable for the outcome of behaviour, or outside intervention, over which one has no authority, or control. To do, or even imply, this will cause irritation and jeopardise co-operation. There are two qualifications to this. First, there are instances in which employees have a legitimate interest in the feedback, even though they may have no direct control, and even little influence, over the performance. An example of this would be feedback of company or project financial performance to junior employees. Second, sensitively handled, peer pressure can be a significant element in promoting behaviour change. Site canteen teasing of under-performers, by their peers, can have more effect than criticism from managers. These issues can impact greatly on the way that performance measures are constructed and used.

Researchers at UMIST have been involved in the design and use of performance measures in construction since 1989. Performance measured has included safety, quality and productivity.

4 SAFETY MEASUREMENT

Safety measurement was addressed, initially, as a fundamental part of a safety improvement initiative, funded by the UK Health & Safety Executive. The essence of the initiative was an attempt to show that safety performance on construction sites could be improved using the behaviourally based technique goal setting and the first objective was 'to develop and test a method of measuring safety performance on construction sites' (Duff et al. 1993; Duff et al. 1994; Phillips 1992).

In most past efforts at safety measurement the emphasis has been on accident frequency. It is a very seductive approach as, after all, the ultimate goal is a reduction in accidents. However, there are serious failings in such data, as a useful performance measure.

The occurrence of accidents is, to some extent, a function of chance in that accidents resulting in injury usually have a number of causes, all of which come together to precipitate the event, and the absence of any of which would have prevented the accident. The severity of an accident is also largely a matter of chance. This all means that it is possible for a safely managed site to have a poor accident performance, and vice versa. Over a long period of time or over a very large site, poor safety management and behaviour will result in a poor accident record; but most construction sites are not very large or very long running. Over a shorter period, a safety performance measure based on accidents can be both unstable, reacting strongly to little change in behaviour, and, at other times, insensitive, failing to react to improvement in behaviour. These facts mean that a safety performance measure, based on accident frequency, is only suitable for large organisations or monitoring over a long time-scale. It does not give the immediacy of feedback that is required for an adequate form of management control.

Another feature of accident reporting is the serious concern about under-reporting, particularly of less serious accidents. The UK Health & Safety Executive estimate this to be of the order of 50% for 'over-3-day' injuries. In addition, the pressure to avoid reporting less serious accidents is increased when they become an element in performance measurement, and more so if the measure is used in appraisal or incentive schemes.

Many of the causes of accidents have their roots in action or inaction well before the event, for example in lack of induction training. This results in considerable time lag between the inappropriate behaviour and any feedback on accident frequency. As a basis for a programme to motivate people to behave safely, it is therefore significantly flawed.

Finally, accidents will only be prevented if people in construction can be persuaded to change their behaviour. This should be the goal; and any measure of performance should focus on issues that are related, as closely as possible, to the aspects of behaviour that require change. Using a medical analogy, focussing on accidents, rather than behaviour, is treating the symptom, rather than the disease.

In summary, accident frequency measures, as measures of safety performance, although quantified, cheaply obtained and objective, are, invalid, unreliable, unstable, insensitive, insufficiently timely and insufficiently related to the behaviour which we try to control. Consequently, it was decided to develop a measure of safety performance that would avoid these deficiencies.

The first part of this process was to identify the major factors in accident causation. As the objective was to develop a measure of safety performance for construction sites, the concern was only with the proximate causes of accidents, those factors that lie within the control of site personnel. A complete list of over 200 causal factors was developed, by reference to safety literature, accident reports and professional experience. This was reduced to 71 by retaining only those that appeared relevant to HSE reported fatal and major injuries. This number was still too large, to include in a measurement process which could be undertaken in a reasonable time, and so the list had to be ranked, in order to identify the most significant items. This could not, unfortunately, be achieved objectively, using accident statistics, as these are rarely classified by the real causes, only the nature of the accident. For example, we know, from HSE reports, how many accidents involve objects falling from height, but not how many involve lack of adequate toe-board protection on scaffolding. For this reason, a questionnaire survey of expert opinion was undertaken, in order to rank the items in terms of their perceived risk in accident causation. 200 site operatives, 200 site managers and 30 contractors safety officers were asked to assess each factor on three scales: frequency of occurrence, likelihood of an accident resulting and typical severity of the likely accident. From 194 responses, the multiple of these factors was used to rank the items. This resulted, initially, in 24 items, classified under four headings, scaffolding, access to heights (other than scaffolding), site housekeeping and personal protective equipment (PPE). A subsequent need to accommodate heavy civil engineering sites resulted in the addition of a plant operation measure. The method of identification and ranking of the items, using a combination of safety literature, reports and expert opinion, ensured high face validity for the measures and there was no adverse opinion expressed, by any of the personnel who were being assessed by the measure, about the validity of the measure.

Although the objective was to frame the measures in terms of items related closely to behaviours, in order to give the initiative a focus on behaviour change, it was not practicable to restrict them to pure behaviour. Most unsafe behaviours are instantaneous and unlikely to be observed during the relatively short measurement process. Using an improperly tied ladder might take some time to observe, but the existence of such a situation in a location clearly intended for access to work in progress would be sufficient evidence of inappropriate behaviour. Similarly, significant amounts of waste materials below scaffolding would be evidence of materials being dropped from the scaffold. It was therefore decided that the measure would focus on unsafe situations, resulting directly from target behaviours, where behaviour could not easily be observed.

The measure pro-forma is supplemented by guidance notes, indicating the issues to be taken into account by the observer, as a set of reminders and indicators in the form 'Look for…'. The score is also supplemented by verbal anchors (descriptors), to be used by less experienced observers. An example of the measure is seen in Figure 1.

Previous behavioural measures of site safety have tended to use an all or nothing approach, total absence of an unsafe situation, within the item being measured, being required to 'score' safe and any infringement resulting in a zero score. This results in a measure that is both unstable, in that it changes readily from zero to 100%, and insensitive, in that it does not respond to smaller changes in performance. To avoid this we designed a proportional rating scale for each item in our measure so that it could be assessed, quantitatively and proportionally. This was achieved by counting that which could readily be counted and sub-dividing the site into sections in order to assess, as objectively as possible, that which could not easily be counted. For example we counted the number of personnel not wearing hard hats, as an item in the PPE measure, and counted the number of walkways, access routes and stairways on the site containing discarded waste materials, under the site housekeeping measure. Although the observations are of the number/proportion of instances of unsafe behaviour/situation, the score is determined by calculating the amount of safe situations or behaviours as a percentage of the total.

It is difficult to assess the sensitivity of the measure objectively. However, during the testing of the measure, in the first phase of the research, a detailed comparison of the recorded scores with any events on site, or in the site environment, that might be expected to influence safety performance, was undertaken. This demonstrated that safety scores reacted strongly to outside interventions. On the occasion of an HSE inspection 'blitz' of the area, scores rose markedly, until such time as a site had been inspected, when the score would fall, presumably on the assumption that 'lightening does not strike twice!' The scores also generally reacted strongly to a site visit by the company CEO.

Housekeeping Category

None					Some					All	Not Seen
0	1	2	3	4	5	6	7	8	9	10	X

1. What proportion of timbers left lying around have nails left in ?

X] Look for nails that have not been taken out or bent over and hammered flat.

X] Look for timbers with protruding nails, left lying around, but ignore timbers that are stacked neatly and/or in skips.

2. What proportion of openings are left uncovered or unguarded ?

X] Look for any opening that is left uncovered or unguarded.

a] Openings in floors should be covered minimum 25mm plywood or have guardrail.

b] Openings in walls should have guardrails, if below waist height (if external) or where adjacent to a drop.

3. What proportion of walkways, access routes and staircases are littered with rubbish or debris ?

X] Look for rubbish or debris on all areas of the site, including all floor levels.

4. What proportion of stored materials are stacked or stored unsafely ?

X] Look for loaded pallets that are stacked more than two high.

X] Look for timbers or steel that are not stacked neatly on 'bites'.

X] Look for scaffold materials that are not stacked neatly.

X] Look for materials that are stacked more than 2.000m (a man's height) high.

X] Look for materials that are stored near to open trenches or excavations.

X] Look to ensure that access gaps are provided between material stacks.

X] Look for overhanging materials obstructing access routes.

X] Look for stacks that are unstable or overhanging.

Figure 1. Sample page of safety measure items.

During the research programme, these measures were used on over 30 sites, ranging from small building projects to motorway construction. The programme was divided into two phases. In the first phase six sites were used to test the performance measures and the effectiveness of goal-setting and feedback as a safety improvement strategy. All the procedures were carried out by the members of the research team. In the second phase (Robertson et al. 1999), the objective was to investigate the practicability of installing the method as part of the safety management procedures of a number of contractors, with the contractors' own personnel carrying out all the measurement, goal-setting and feedback procedures.

In order to establish the reliability of the measure, inter-observer reliability checks were carried out. In the second phase, these involved repeated occasions on which a member of the research team would 'shadow' the contractors own observer and produce an independent safety measure. The recorded scores were then compared, on an item by item basis. The worst inter-observer reliability score recorded, for any site, was 92.1% and 15 of the 24 sites tested recorded better than 96% agreement. The measure was, when used by a number of different observers, therefore, very reliable. Reliability was also assured, in part, by carrying out observations frequently, but at varying times and using observers who had additional tasks, thus making it difficult for operatives to anticipate observations.

The economic viability of the measure is best assessed by comparing the induction and operation times, for the observers, with the benefits that might be obtained from its use. On a typical, confined building site, such as an office block or hospital extension, the observer took 45-60 minutes to take each complete site safety measurement. For the purposes of the goal-setting and feedback programme, this was usually repeated three times each week. On the motorway project, the duration of the measurement process was approximately half-a-day, but involved a round trip of over 40 miles and frequently included other duties such as transmitting engineering information. On this site, the frequency was reduced to twice per week.

The training time for an observer was two days, of which only half was related to the use of the measure; the rest was training in goal-setting and feedback procedures. The training cost could, of course, be distributed over many sites as the observer would be competent to introduce the process on subsequent projects.

The 'cost' for introducing the measure on a typical building site would, therefore, be less than one man-days training plus 2-3 man-hours per week, including calculation and recording time. The benefits will depend on the use to which the measure is put. In this project, the measure was used as the basis of a goal-setting and feedback programme which resulted in a mean reduction in unsafe behaviour of 23%, over 15 construction sites. A recent study by HSE suggested that the real cost of injurious accidents could be 1% of construction cost so, were this safety improvement to be translated into an equivalent reduction in accident frequency, then the economic benefit could outweigh the cost by approximately 20 times.

The measure was designed to include only those items that are high risk, relate to general construction activity and occur frequently. By design, therefore, the measure is valid for most construction sites and over most of the period of construction. It is very flexible, in that it will cope with sites that include only some of the items or classes. It can, for example, be used on sites without any heavy plant, or scaffold. It can also be used on occasions when some items are 'unseen'. In these circumstances, of course, comparisons of one site with another have to be made with care.

On the question of timeliness, the results of the measure can be available on the day of measurement. In the project described here, the score used in the feedback was calculated as a weekly average, usually of three sets of observations taken at varying times, and posted on feedback charts each Monday morning. Thus, the 'reward' from improved performance was received within a day or two.

A qualitative observation of the effect of management commitment on the success of this safety improvement intervention, on the sites observed during the first phase of the research, led to the development of an objective measure of management commitment to use in the second phase. It had been observed that the success of the intervention, measured as the proportional improvement in safety behaviour, seemed to be correlated with the apparent commitment of the site management to the intervention process. This prompted the research team to attempt to test this hypothesis in the second phase. A measure of management commitment was designed, to include the following items, all assessed on a scale of 1 to 5: 'Attitude to placement of feedback charts'; 'Attitude to workers stopping work to attend goal-setting meetings'; 'Proportion of management team attending goal-setting meetings and acting supportively'; 'Attitude to observer taking time away from other duties, to perform the measure'; 'Overall rating given by the observer during a research project de-briefing'. The use of this measure demonstrated clearly that management commitment was having a highly statistically significant effect on the effectiveness of the intervention. The overall improvement in safety performance, measured by the effect size indicator (Cohen's 'd'), had a coefficient of correlation, r, of 0.55 with the management commitment measure, at a 1% significance level.

The same approach to safety performance measurement, as developed for this HSE sponsored project, has been applied in two, more specialised situations. The first of these was a measure developed, for one of the contractors supporting the initial project, to cope with high-risk aspects of plant operation adjacent to live motorway carriageways. The potential for serious accidents to the motoring public involving plant operating on motorway widening contracts persuaded the contractor that a measure was required which contained items derived from all the elements of good practice in operating plant, particularly in confined conditions. The measure included items such as correct application of machine stabilisers, use of banksmen and positioning of machine in relation to traffic barriers, boom reach and counterweight rotation.

The second situation involved developing an extended PPE measure for a manufacturer of such equipment to measure performance in the maintenance and use of ear defenders and respirators. The major extension to the normal PPE usage measure was to include the selection of the correct equipment, correct wearing of the equipment and equipment maintenance.

5 QUALITY MEASUREMENT

Before discussing the work carried out at UMIST on construction quality measurement, it is helpful to recognise the distinction between quality of production process and quality of product. Quality of process relates to the physical and organisational procedures that lead to the delivery of the product or service. Some processes are not carried out by people but, when they are, then the quality of the process is a function of the behaviour of the operative, supervisor or management. Quality of product relates to the characteristics of the output of the process. The quality of the process will influence the quality of the product but it will also influence many other outcomes, such as levels of wastage or environmental pollution. The quality of the product will be influenced by the quality of the process but it is also influenced by other factors, such as the quality of the materials or work environment. In this research we were concerned, predominantly, with the quality of the product but, in seeking a measure which exercises effective control over the quality of the product, it was necessary to focus on the factors which are most influential in achieving good quality. These are often factors in the production process, particularly the behaviour of the operatives. This, of course, as we have seen with safety measurement, presents a problem. Systematically repeated behaviour is observable. Instantaneous behaviour is not, at least without excessive observation effort. The solution is, therefore, to observe the consequences of the behaviour, usually features of the output or product. However, this also causes problems. First, the consequences are frequently hidden, buried or otherwise indistinguishable. Consider the porosity of concrete due to inadequate vibration, or unfilled joints in masonry. The solution adopted, therefore, must take all these factors into account and the following principles are followed in the development of the quality measures:

- As far as possible, measure behaviour or its immediate outcome, that is focus on the things that may need improving;
- If behaviour is systematic, that is regularly repeated, measure behaviour;
- If behaviour is infrequent or irregular, and difficult to observe, then measure the outcome that is visible, as soon after the behaviour as possible.

6 CONCRETE QUALITY MEASURE

The measure of concrete quality was developed under contract for a major UK civil engineering company (Phillips & Duff 1997) and tested on a large coastal engineering project, with reinforced concrete pours of several hundred cubic metres, and on aircraft parking aprons at a large international airport.

The main objectives of the project were to:

- Determine appropriate categories of performance and items to include in the measure;
- Design scales of measurement, including appropriate weighting, for each item and category;
- Test the measure for face validity through consultation with company personnel.

Once again, a literature review was the basis of a comprehensive list of observable behaviours and situations relating to concrete workmanship and causes of defects. The list was classified under five headings:

- Pour planning and preparation;
- Formwork and falsework;
- Reinforcement;
- Concrete discharge, transport and placing;
- Concrete finishing and protection.

Company materials engineers then assessed the perceived risk of each item on scales of frequency of occurrence (behavioural failure), likelihood of defect resulting and severity or 'cost' of the consequences.

In contrast to the results of a similar procedure in the development of the safety measure, the result of this did not permit discrimination between items in order to identify the most important, or high risk, items. All items were perceived to be important. This could have caused a serious problem in overloading the observer. However, as the items were classified under headings related to a sequence of concreting operations, carried out at different times, the problem was avoided. None of the categories of measure contained too many items to be recorded, at one time, by the observer.

In the case of some of the items, an all or nothing approach was necessary. For example, there can be no discretion in assessment of reinforcing steel grade; it is either right or wrong. Whereas, in other items, there is more discretion and a degree of subjectivity may be permitted. The quality of the tamped finish of a concrete slab was recorded as one of the following: 'perfect'; 'some evidence of poor tamping, but surface does not cause serious concern'; or, 'much evidence of incorrect tamping, may require remedial work'.

In order to emphasise the importance of compliance with specification, it was decided to distort the measurement scale by weighting non-compliance more heavily than compliance. In the score evaluation, any item assessed as possibly requiring remedial work or replacement was given a score of *minus* five, whereas items assessed as compliant were given a score of one, two or three, depending on the level of perceived quality above the compliance threshold. These weightings make the measure very sensitive to compliance but can be modified to reflect the perceived seriousness of non-compliance; and there may well be a case for different negative weightings, depending on the potential cost of remedial work. An example of the measure is seen in Figure 2.

In the test of the measure on airport 'pavement quality' concrete, the form of construction was un-reinforced flat slabs. This meant that only the sections of the measure directly relating to concrete could be used; formwork and reinforcement items were excluded. The test involved eight uses of the measure, over a three-day period, by a researcher, a materials engineer and site supervision. The level of control of the construction process was very high, the work repetitive and the concrete trades gangs very experienced. The result was very high scoring. The inter-observer reliability checks gave a percentage agreement score of 97%, probably affected by the ceiling effect reducing the variability of the scores, with the only discrepancies between observers occurring in the items of measurement focussing specifically on operative behaviour.

The very high scores already being achieved meant that use of the measure was unlikely to significantly improve the quality of the process. However, the materials engineer considered that regular, weekly/fortnightly use of the measure would be a useful addition to the normal quality management procedures.

In the coastal engineering project the whole scope of the measure was tested over two days by a researcher, a materials engineer and, for some of the categories, by site engineering staff. The size of the concrete pours, the one observed being approximately 800 cubic metres, raised some interesting issues over the timing of the measurement process. In smaller pours, the measures have a clear sequence: pour planning and preparation; concrete discharge, transport and placing; finishing and protection. However, in a very large pour all three operations can be continuing at the same time, making the timing of the individual categories of the measure difficult. The result is that the measure becomes, in part, a check-list for continuing supervisory control, rather than a measure of the outcome of the completed process. The company is considering development of the measure specifically for this purpose.

Some concern was expressed about the 'tick-box' format of the measure which, although convenient, might encourage less rigorous attention and careless, unreliable, use. The inter-observer reliability test gave a result of 81% agreement, with the discrepancies largely between the items requiring an assessment of *degree* of compliance. For example,

ITEMS DESCRIPTION	%	not seen
10. What proportion of **cavity** is formed inadequately?	☐	[ ]

a) Look for blockage in the cavity of the cavity walling created by protrusions of mortar, broken bricks and DPC.
b) If width of cavity is not specified, it should not be less than 50mm in general
c) In general, cavity should be cleaned and cleared of mortar droppings

11. What proportion of the **wall ties** are fixed incorrectly and/or inadequately?	☐	[ ]

a) Look for fixed wall ties that sloped towards the inner leaf of the cavity wall
b) Look for patches of wet spots occurred on the inner leaf of the wall.
c) In general, wall ties should be bedded equally into both leaves of the cavity wall or half thickness of each leaf of the wall.
d) In general, there should not be less than 5 ties in any one square of the cavity walling.
e) In general, the spacing of wall ties should not exceed 600mm apart horizontally and 450mm vertically and they should staggered on alternate layers to sit central to the previous ones.
f) Spacing of wall ties can be referred to BS 8000: Part 3: 1989 Table 3 (refer Attachment F). Guidance to spacing of additional ties at sides of openings can be obtained from BS 8000: Part 3: 1989 Figure 6 (refer Attachment F).

12. What proportion of the **insulation** in the cavity wall is not fixed properly?	☐	[ ]

a) In general, insulation boards should be secured by staggering the vertical joints of the board and the joints should butt closely. Do not overlapped the boards except in the case of those having specially rebated edge.
b) In general, with full-fill or no cavity insulation, there should be a minimum of 50mm cavity between leaves. With partial-fill cavity insulation, a minimum of 50mm residual air space is recommended.
c) In general, two wall ties should be provided at the top and bottom of each batt of insulation; at least 600mm apart horizontally.
d) In general insulation should not tear or force over the extra ties at window and door jambs
e) In general, horizontal joints of insulation should correspond with the horizontal rows of ties
f) Small strips cut from the batt of the insulation should not lean outward and bridge the cavity of the cavity wall.
g) Insulation should be built up to verge of the gable cavity wall if cavity trap is not allowed.
h) For additional information, reference can be made to Item 3.4.5 Insulation of Attachment F and Attachment G.

13. What proportion of the **cavity traps** have gaps, perforations and there are no effective stop ends?	☐	[ ]

a) Look for damp patches on inner leaf of the cavity wall, particularly below the cavity tray.
b) Look for cross joints immediately above cavity trays that are left open as weep holes. Weep holes should be placed at not more than 1m centres (every fourth cross joints) but at least two above any opening.

Figure 2. Sample page of concrete quality measure.

on occasion, one observer would score an item as compliant, but imperfect, whereas the other observer would score the item as non-compliant. If the items had been simply scored as compliant or non-compliant, an all-or-nothing approach, then the inter-observer reliability score would have been 92%.

At both sites, the measure was well received, seen to be valid and a potentially useful management tool, given some slight modifications to format and careful attention to the timing of its use on large pours. The reliability on the large pour was not high and requires improvement. It is believed that this would happen as experience increased, in particular in reaching agreement about the thresholds of performance, when scoring items with graded levels of compliance.

7 BRICKWORK QUALITY MEASURE

The measure of brickwork quality (Tang, 1997) was developed in a similar way to the concrete measure. The literature was searched to identify appropriate items, which were classified under the following categories:
- Handling and site storage of materials;
- Planning and preparation of work;
- Appearance;
- Jointing and pointing;
- Damp-proof course;
- Cavity walling;
- Coping and capping;
- Movement joints;
- Fixing lintels;
- Protection to brickwork.

In the case of the brickwork quality measure, all the items related to situations, potential quality faults, resulting directly from incorrect behaviour on the part of site supervisors or operatives. Eight people, experienced in brickwork, were asked to rate each item a verbally-anchored numerical scale [never (0); sometimes (1); often (2); always (4)] of the frequency of occurrence; and on a similar scale of potential cost resulting from the fault. The people included a brickwork foreman, a bricklayer, 2 brickwork tutors, a builder and 3 clerks of works. The items were then ranked on the multiple of the two measures. The original list contained 100 items, far too many to include in a usable measure, so two methods were used to reduce this number. First, any items scoring less than the mean score of all the items were rejected. This reduced the number to 37. Second, the remaining items were collapsed, where possible, by combining similar items, in each category, into composite items, thus reducing the eventual measure to 20 items. Each item is subject to a measure that requires the observer to assess the proportion of the observed work that is defective, as a percentage of the total. Each item in the measure is supplemented by a number of guidance notes, of the form 'Look for…'. These could be removed, in order to make the pro-forma more succinct, once the observer is thoroughly familiar with the procedures, though care would be needed to ensure that this did not lead to excessive subjectivity in the judgements of degree of compliance with good practice. The total score is calculated as a mean value, with all items given equal weighting. Figure 3 shows a section of the measure.

Preliminary testing of the measure has been carried out on two sites, one a residential development and the other a commercial office development. In the first case the observations were carried out by the contractor's Site Agent (manager), and, in the other, by the client's Clerk of Works. Measure development is not yet complete and reliability tests have not yet been possible. However, experience and comments from the observers, so far, suggest that the measure is:

- Appropriate for large or small sites;
- Takes approximately 30-45 minutes to use, depending on the degree of dispersion of the bricklaying activity, and is therefore economically viable;
- Is easy to learn to use, providing the observer has technical experience;
- Would be capable of measuring changes in quality over time, on the same site;
- Could be used for comparison of different sites, subject to caution if the items observed are different on the observed sites.

Item number	CONCRETE TRANSPORT, DISCHARGE AND PLACING MEASURE:	Compliance	Non-Compliance
3.4	Supply breakdown, if relevant.		
	Inadequate response, causing serious actual or potential technical problems		5
	Adequate response, but did not operate agreed contingency plan	1	
	Followed contingency plan, which worked	2	
	Contingency plan did not work	0	
3.5	Weather plans, if relevant.		
	Inadequate response, causing serious actual or potential technical problems		5
	Adequate response, but did not operate agreed contingency plan	1	
	Followed contingency plan, which worked	2	
	Contingency plan did not work	0	
3.6	Safety - safety plan.		
	Safety plan ignored		5
	Safety plan generally followed, but not in every respect	1	
	Safety plan followed in every respect	2	
3.7	Safety - PPE.		
	Several instances of correct PPE not worn		5
	A few (one or two) instances of PPE not worn	1	
	No instances of PPE not worn	2	
Page Quality Score:		[illegible]	[illegible]

Total Compliance - Total Non - Compliance

☐ - ☐ = ☐

Transfer Quality Score to page 5

Page 3

Figure 3. Section of brickwork quality measure.

ACKNOWLEDGEMENTS

This chapter contains ideas and information derived from a number of research projects to which a large number of people have contributed. I would particularly like to record my thanks and appreciation to my research colleagues at UMIST: Professor Ivan Robertson; Robin Phillips; Dr. Dominic Cooper; Dr. Tim Marsh; Sy Bor Tang; and Andy Weyman of the UK Health & Safety Executive.

We all also owe a considerable debt of gratitude to the large numbers of managers, engineers, supervisors and operatives who allowed us unfettered access to over 40 construction and industrial sites. They are far too numerous to mention by name but it should be recorded that they gave their time, and much welcome comment and advice, so willingly, in support of our attempts to develop performance measurement tools that would contribute to their safety and quality objectives. The following companies were represented: Alfred McAlpine Construction Ltd., Balfour Beatty Construction Ltd., Balfour Beatty Civil Engineering Ltd., British Steel Ltd., Bovis Construction Ltd., Costain Construction Ltd., Courtaulds Chemicals Ltd., Harmann and Reimer Ltd., Higgs and Hill (Northern) Ltd., John Laing plc., Kvaerner Oil and Gas Ltd., Lovells Ltd., Morrison Construction Ltd., Norwest Holst Group Ltd., Tarmac Construction Ltd., Taylor Woodrow Ltd., 3M Ltd., Trafalgar House (Regions) Ltd. (now Kvaerner Construction Ltd.)., Wimpey Construction Ltd.

Notwithstanding my reliance on the efforts of others in writing this chapter, I take full responsibility for the accuracy of its content and the way that it represents the ideas contained herein.

REFERENCES

Duff, A.R., Robertson, I.T., Cooper, M.D. & Phillips, R.A. 1993. Improving safety on construction sites by changing personnel behaviour, HSE Report Series CRR 51/93, H.M.S.O., London, ISBN 011882 1482.

Duff, A.R., Robertson, I.T., Cooper, M.D. & Phillips, R.A. 1994. 'Improving safety by the modification of behaviour', *Construction Management and Economics*, 12: 67-78, E. & F.N. Spon.

Phillips, R.A. 1992. The development of a safety performance measure for the construction industry, MSc Thesis (unpublished), UMIST (University of Manchester Institute of Science and technology), England.

Phillips, R.A. & Duff, A.R. 1997. Quality Measurement on Construction Sites report prepared for BICC Group Research Consultancy Scheme (unavailable without the express permission of BICC/Balfour Beatty Civil Engineering Ltd.)

Robertson, I.T., Duff, A.R., Marsh, T.W., Phillips, R.A., Weyman, A.K. & Cooper, M.D. 1999. Improving safety on construction sites by changing personnel behaviour – Phase Two, HSE Report (in print – to be published in 1999).

Tang, S.B. 1997. Development of a Brickwork Quality Measure, MSc Dissertation (unpublished), UMIST (University of Manchester Institute of Science and Technology), England.

CHAPTER 2

Management of safety, health and the environment in construction

LUÍS M. ALVES DIAS
Instituto Superior Técnico, Department of Civil Engineering, Lisbon, Portugal

ABSTRACT: Safety and health of workers and the protection of the environment have been of great concern during recent years to many parties involved in the construction process in many countries around the world. In most cases, a significant number of laws and regulations were enacted by these countries, driven by recommendations of international organizations, such as the International Labor Organization, World Health Organization, International Organization for Standardization, among others. Other countries are following general directives, which are designed to be applied by all members of a group, as with the European Union.

In any case, these regulations reflect the world concern on the need to reduce the high number of occupational accidents and illness attributed to the workplace and to reduce pollution. These regulations are, in general, based on the technological knowledge that is being developed by experts on each specific matter.

The practical and successful implementation of the requirements of the new laws and regulations, requires management approaches that vary according to the specificity of the industry for which they must be applied. In the construction industry, the implementation of safety and health on construction sites has been applied, for many years, in different ways based on the knowledge and on the regulations existing in each country. Implementation of environmental agenda has not been addressed on a regular basis on most construction sites. Recently, in some countries, construction companies are beginning to develop their own systems on a voluntary basis.

In this chapter, an approach to the implementation of systematic and integrated occupational safety and health and environmental management systems on construction sites is presented.

1 INTRODUCTION

Since the 1987 publication of the first international quality management standards (ISO 9000) by the International Organization for Standardization (ISO), many construction companies worldwide have been implementing their own quality systems driven mainly by their client's demands or market pressures. In some cases, they were driven by the recognition of the profits that they could realize through better organization.

Preventive measures for the reduction of occupational injuries and illness on construction sites has been implemented for many years by many construction companies driven by the applicable regulations, with governmental inspectors having the role of verifying the application of these regulations. This official inspection is usually casual due to the reduced number of inspectors in each country.

Some construction companies have been implementing prevention measures on construction sites driven by the recognition of the value of their human resources. More recently, some construction companies are making efforts to address safety and health in their quality systems, by developing specific procedures for these objectives (inspection and prevention plans against falls). Others are developing their own safety systems, using guides already existing in such countries as the UK and Australia.

Reduction of the environmental impacts of construction sites (noise, dust, etc.) is also considered of great value by some construction companies, although in this case they are driven in general by requirements of clients that want to avoid neighborhood and/or public claims. In some cases, however, some construction companies are developing environmental management systems following the ISO 14000 standards. Other companies, recognizing the high costs of implementing environmental programs as a separated issue, are integrating them with occupational safety and health programs.

It is the author's belief that in the beginning of this millennium, construction companies will recognize the benefits (on costs and time for implementation) of integrating environmental, quality and occupational safety management systems, leading to the development of one single system – EQS management system.

2 MANAGEMENT SYSTEM STANDARDS

The International Organization for Standardization (ISO[1]) is a private and non-governmental organization. Since post World War II, ISO has been developing standards on the basis of international consensus, focusing almost exclusively on products. Since then, more than 11000 international standards have been published by ISO covering many technical fields.

During the 1980s, for the first time in its history, ISO began the task of creating standards that did not have a technical basis but dealt with organizational management – the ISO 9000 standards for quality management. The first series of these standards was issued in 1987. Table 1 lists the actual family of these standards, some of which (particularly, italic ones) have also been used as guidelines to developing documentation for occupational safety and health management systems.

2.1 *Environmental management standards*

The success of the ISO 9000 standards, adopted worldwide by many organizations as a requirement for trade, has driven ISO to develop other organizational standards. In the early 1990s, ISO started the work on standardization in the field of environmental management – the ISO 14000 standards. The first series of these standards were issued in

---

[1] ISO, is a Greek word meaning *equal*.

Table 1. List of ISO 9000 family of standards.

	Concepts and terminology
ISO 8402:1994	Quality management and quality assurance – Vocabulary
	Quality systems
ISO 9000-1:1994	Quality management and quality assurance standards – Part 1: Guidelines for selection and use
ISO 9000-2:1997	Quality management and quality assurance standards – Part 2: Generic guidelines for the application of ISO 9001, ISO 9002 and ISO 9003
ISO 9000-3:1997	Quality management and quality assurance standards – Part 3: Guidelines for the application of ISO 9001:1994 to the development, supply, installation and maintenance of computer software
ISO 9000-4:1993	Quality management and quality assurance standards – Part 4: Guide to dependability programme management
ISO 9001:1994	Quality systems – Model for quality assurance in design, development, production, installation and servicing
ISO 9002:1994	Quality systems – Model for quality assurance in production, installation and servicing
ISO 9003:1994	Quality systems – Model for quality assurance in final inspection and test
ISO 9004-1:1994	Quality management and quality system elements – Part 1: Guidelines
ISO 9004-2:1991	Quality management and quality system elements – Part 2: Guidelines for services
ISO 9004-3:1993	Quality management and quality system elements – Part 3: Guidelines for processed materials
ISO 9004-4:1993	Quality management and quality system elements – Part 4: Guidelines for quality improvement
ISO 10005:1995	*Quality management – Guidelines for quality plans*
ISO 10006:1997	Quality management – Guidelines to quality in project management
ISO 10007:1995	Quality management – Guidelines for configuration management
ISO 10011-1:1990	*Guidelines for auditing quality systems – Part 1: Auditing*
ISO 10011-2:1991	Guidelines for auditing quality systems – Part 2: Qualification criteria for quality systems auditors
ISO 10011-3:1991	Guidelines for auditing quality systems – Part 3: Management of audit programs
	Supporting technologies
ISO 10012-1:1992	Quality assurance requirements for measuring equipment – Part 1: Metrological confirmation system for measuring equipment
ISO 10012-2:1997	Quality assurance for measuring equipment – Part 2: Guidelines for control of measurement processes
ISO 10013:1995	*Guidelines for developing quality manuals*
ISO/FDIS 10014	Guidelines for managing the economics of quality
ISO/DIS 10015	Quality management – Guidelines for training

1996. Figure 1 shows the organization of the actual family of ISO 14000 standards. These standards are listed in Table 2.

With respect to quality, ISO 14000 provides organizations with guidelines to develop effective environmental management systems. They address the process that the organization uses to create a product. They do not address the environmental characteristics of the product, i.e., it is not a 'green' or 'environmentally friendly' product label.

Although all those standards have value and are interconnected, for certification purposes an organization seeking an 'ISO 14000 certificate' must only show that its envi-

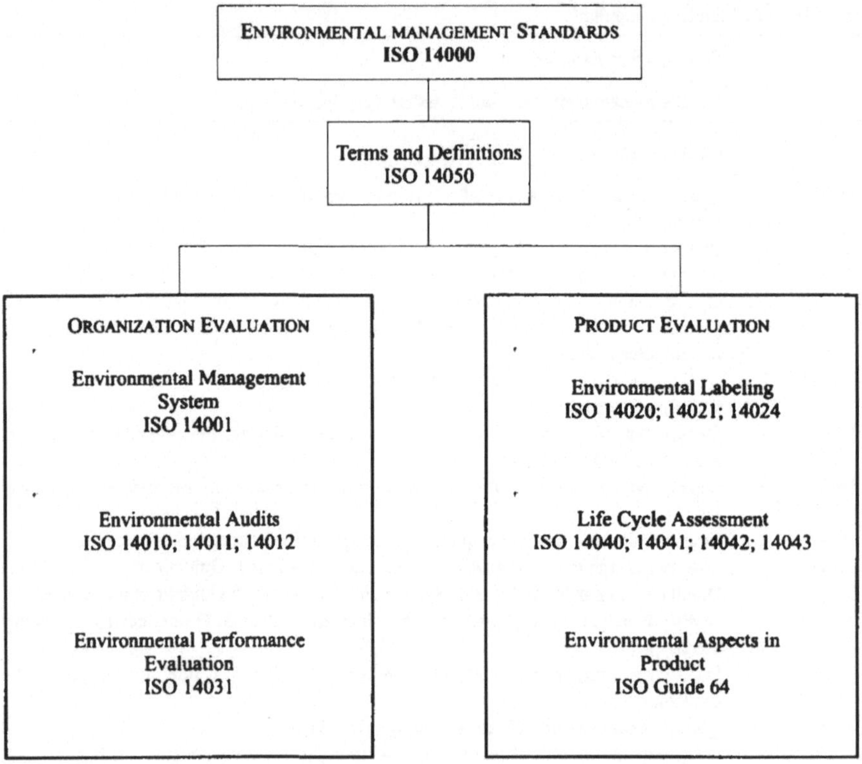

Figure 1. ISO 14000 Standards Organization.

ronmental management system is in conformance with the requirements of ISO 14001. A certification body must verify this conformance after auditing the system. The main elements of ISO 14001 are shown in Figure 2.

General Requirements may include a generic description of the management system (its documentation and resources), which will be dependent on the size of the organization and the nature of its activities. It must also specify whether it applies to the entire organization or to some of its activities which must be identified.

Environmental Policy must be defined and adopted by top management, be documented, implemented, maintained and communicated to all employees, and be made available to the public and other interested parties. It must include at least the following statements:

1. Commitment to continual improvement and prevention of pollution;
2. Commitment to compliance with relevant environmental legislation and regulations;
3. Provide for the framework of setting and reviewing environmental objectives and targets.

Planning must include:

1. A procedure to identify the environmental aspects of the project activities (waste generation, wastewater discharge, stormwater discharge, etc.) and determine which of those can have a significant impact on the environment (impacts on ecology, on natural resources, on pollution, etc.);

Table 2. List of ISO 14000 family of standards.

	Environmental management systems
ISO 14001:1996	Environmental Management Systems – Specification with Guidance for Use
ISO 14004:1996	Environmental Management Systems – Guidelines on Principles, Systems and Supporting Techniques
	Environmental auditing and related environmental investigations
ISO 14010:1996	Guidelines for Environmental Auditing – General principles on environmental auditing
ISO 14011:1996	Guidelines for Environmental Auditing – Audit Procedures – Auditing of environmental management systems
ISO 14012:1996	Guidelines for environmental auditing – Qualification criteria for environmental auditors
ISO/WD 14015	Environmental Assessments of Sites and Entities
	Environmental labeling
ISO/DIS 14020	Environmental Labels and Declarations – General principles
ISO/DIS 14021	Environmental Labels and Declarations – Environmental Labeling – Self-declaration Environmental Claims
ISO/DIS 14024	Environmental Labels and Declarations – Environmental Labeling Type I – Guiding principles and procedures
ISO/WD 14025	Environmental Labels and Declarations – Environmental Labeling Type III – Guiding principles and procedures
	Environmental performance evaluation
ISO/DIS 14031	Environmental Performance Evaluation
	Life cycle assessment
ISO 14040:1997	Environmental Management – Life Cycle Assessment- Principles and Framework
ISO/FDIS 14041	Life Cycle Assessment – Life Cycle Inventory Analysis
ISO/CD 14042	Life Cycle Assessment – Impact Assessment
ISO/CD 14043	Life Cycle Assessment – Interpretation
ISO/NP 14049	Life Cycle Assessment – Examples for the application of ISO 14041
	Terms and definitions
ISO/FDIS 14050	Environmental Management – Vocabulary
	Environmental aspects in product standards
ISO Guide 64	Guide for Environmental Aspects in Product Standards

2. A procedure to identify and access legal and other requirements to which the organization subscribes;
3. A procedure to establish objectives (reduce effluents and emissions, reduce energy consumption, reduce water consumption, increase product recycling, etc.) and targets (20% reduction in wastewater discharge next year, reuse or recycle wood pallets next year, etc.);
4. A procedure to establish and maintain an environmental program assigning responsibilities and means for achieving the stated objectives and targets.

Implementation and operation must include:

1. Development of the organizational structure and assignment of responsibilities;
2. Establishment of a program for training, awareness and competence for all employees – also for contractors and suppliers, whose work can impact the environment;

Figure 2. Environmental management system elements (ISO 14001).

3. Establishment of a program for internal and external communication on all environmental issues with employees, neighbors, the general public and customers;
4. A list of all management system documentation;
5. A procedure for control and keeping current the documents of the management system (in paper or electronic form) in a way that ensures obsolete documents are not used;
6. A procedure for operational control to identify the organization's operations and activities associated with significant environmental aspects; and
7. A procedure for emergency preparedness and response.

Checking and corrective action must include:

1. A program for monitoring and measuring the system to assess the progress of the organization in meeting objectives and targets;
2. A procedure for investigating and correcting nonconformances (deviations from requirements of the standard) and to prevent recurrence of the same kind of nonconformance;
3. A procedure to maintain environmental records which will allow the organization to demonstrate conformance to the standard (information pertaining to environmental laws, training, audit results, incident reports, etc.); and
4. An audit program to determine whether the environmental system conforms to the requirements and is properly implemented and maintained.

Management review is important to ensure that the system is suitable, adequate and effective. It must be performed and documented at defined intervals and includes reviewing the results of audits, and the assessing the initial objectives to assure that targets are being met.

The organization of these elements follows the Plan-Do-Check-Act process, including the environmental policy and planning (Plan), the implementation and operation (Do), the checking and corrective action (Check) and the management review (Act).

2.2 *Occupational safety and health management standards*

Following the ISO 9000 and ISO 14000 series of standards, a discussion on the need for the development of an international standard for occupational safety and health management system emerged during 1996. Many meetings were organized to discuss this matter; the most important being the workshop sponsored by ISO in September 1996. The main conclusion of this workshop was not to support, the ISO development of an international standard for safety and health at that time (ANSI, 1996). Considering this recommendation, the Technical Management Board (TMB) of ISO decided that no further action should be taken, to initiate activity within ISO in the field for such a standard.

The reasons for this decision were: insufficient experience to evaluate the benefits and effectiveness of the quality and environmental management standards; possible conflict with existing national laws and regulations concerning safety and health; and employer-labor relations. It was also considered that such a management system should be developed at a national or regional level, due to different socio-economic and cultural conditions.

Reasons given in favor of the development of such an international standard included: improvement of safety and health; enhancement of business performance; provision for integration into quality management systems; and help in starting the process of improvement in countries where regulations do not yet exist or are not sufficiently developed.

At the same time ISO/TMB decided not to initiate the development of an international standard on safety management system, it was also decided to study the possible integration of the ISO 9000 (quality) and ISO 14000 (environment) series of standards (ISO, 1997). The main recommendation of a technical advisory group was not to merge the two families of standards, but to make them more compatible and easier to implement. For this group the term *compatibility* 'means that common elements of the standards can be implemented in a shared manner, in whole or in part, by organizations without unnecessary duplication or the imposition of conflicting requirements'. This compatibility between the two families of standards may start with the standards for quality and environmental auditing. A discussion for developing general principles for management system standards is also being considered. This could facilitate compatibility and also the integration of occupational safety and health issues.

Meanwhile, some countries, like the United Kingdom, started publishing guides for occupational safety management systems such as the standard BS 8800 – Guide to occupational health and safety management systems (BSI, 1996). This standard offers two different approaches. One is based on the ISO 14001 standard using the same elements with necessary adaptations. The other is based on the elements shown in Figure 3.

2.3 *Other approaches related to occupational safety and health management systems*

The American Industrial Hygiene Association (AIHA, 1996b) also published a guidance document for the development of occupational health and safety (OHS) management systems. Based on the ISO 9000 series standards, they proposed the elements shown in Table 3.

3 INTEGRATION OF MANAGEMENT SYSTEMS

In a recent work conducted by the Department of Civil Engineering of Instituto Superior Técnico (Alves Dias et al. 1998), a possible integration of the quality and occupational safety and health systems (Q&S) was explored. Table 4 shows the main elements of such

Figure 3. Occupational safety management system elements (BS 8800).

Table 3. Safety management system elements (AIHA).

1. OHS management responsibility
2. OHS management system
3. OHS compliance and conformance review
4. OHS design control
5. OHS document and data control
6. Purchasing
7. OHS communication systems
8. OHS hazard identification and traceability
9. Process control for OHS
10. OHS inspection and evaluation
11. Control of OHS inspection, measuring and test equipment
12. OHS inspection and evaluation status
13. Control of nonconforming processes or devices
14. OHS corrective and preventive action
15. Handling, storage and packaging of hazardous materials
16. Control of OHS records
17. Internal OHS management system audits
18. OHS training
19. Operations and maintenance services
20. Statistical techniques

an integrated system. The italic lines represent the elements of ISO 9001 whose designations were adapted to address occupational safety and health issues.

Table 5 shows a comparison between these elements (representing a possible integration of quality and occupational safety management systems) and the elements of ISO 14001 (environmental management systems).

It is enlightening that most of the elements of ISO 14001 may be integrated with one or more elements of the proposed integrated system. Only four of the elements of ISO 14001 have no direct correspondence to elements of this integrated system (4.3.1 – environmental aspects; 4.3.4 – environmental management program; 4.4.3 – Communication; 4.4.7 – emergency preparedness and response).

Table 4. Elements of a possible integration of a Q&S system.

1. Management responsibility
2. *Quality and occupational safety management system*
3. Contract review
4. Design control
5. Document and data control
6. Purchasing
7. Control of customer-supplied product
8. Product identification and traceability
9. Process control
10. *Inspection, testing and hazard prevention*
11. Control of inspection, measuring and test equipment
12. Inspection and test status
13. Control of nonconforming product
14. Corrective and preventive action
15. Handling, storage, packaging, preservation and delivery
16. *Control of quality and occupational safety records*
17. *Internal quality and occupational safety audits*
18. Training
19. Servicing
20. Statistical techniques

For these elements, a possible relation may be recognized to the elements designated with an open circle (4.10 – Inspection and testing for 4.3.1 and 4.4.7; 4.2 – Quality and occupational safety systems for 4.3.4; 4.5 – Document and data control for 4.4.3). Another approach is to add to the list of the twenty elements of the Q&S integrated system. Presently a study is under development at IST (Technical University of Lisbon) to explore this issue to greater degree.

It is the author's belief that integration of environmental, quality and occupational safety and health management systems can result in greater cost effectiveness and time reduction for implementation.

As shown in Table 5, some requirements of both standards (ISO 9000 and ISO 14000) are common, or can be shared, adapted and/or complemented. Integration will also benefit each of the three areas on focus (E, Q and S) by the effect of 'inducing', i.e., each may benefit from the experience of the implementation of the other. This integration of the management systems is shown in Figure 4.

4 STRUCTURE OF AN EQS MANAGEMENT SYSTEM

The EQS management system can be structured and documented, according to Figure 5, in the following categories:

- EQS management Manual;
- EQS management general Procedures;
- EQS management Plans;
- Work Instructions or Procedure, and Inspection, test and risk prevention Procedures.

It must be emphasized that this structure may be used for the implementation of management systems in an organization that decides to involve all three areas (EQS), or

Table 5. Elements of the proposed integrated Q&S system versus those of ISO 14001.

Environmental management system requirements (ISO 14001) / Quality and occupational safety systems requirements (ISO 9001 adapted)		4.1	4.2	4.3 Planning 4.3.1	4.3.2	4.3.3	4.3.4	4.4 Implementation and operation 4.4.1	4.4.2	4.4.3	4.4.4	4.4.5	4.4.6	4.4.7	5 Checking and corrective actio 4.5.1	4.5.2	4.5.3	4.5.4	4.6
		General requirements	Environmental policy	Environmental aspects	Legal and other requirements	Objectives and targets	Environmental management programme	Structure and responsibility	Training, awareness and competence	Communication	Environmental management system documentation	Document control	Operational control	Emergency preparedness and response	Monitoring and measurement	Nonconformance and corrective and preventive action	Records	Environmental management system audit	Management review
4.1	Management responsibility		●			●		●											●
4.2	Quality and Occupational Safety Systems	●					O				●		●						
4.3	Contract review																		
4.4	Design control				●								●						
4.5	Document and data control									O		●							
4.6	Purchasing												●						
4.7	Control of customer supplied product												●						
4.8	Product identification and traceability																		
4.9	Process control												●						
4.10	Inspection and testing			O										O	●				
4.11	Control of inspection, measuring and test equipment														●				
4.12	Inspection and test status																		
4.13	Control of nonconforming product															●			
4.14	Corrective and prevention action															●			
4.15	Handling, storage, packaging, preservation and delivery												●						
4.16	Control of quality and occupational safety records																●		
4.17	Internal quality and occupational safety audits																	●	
4.18	Training								●										
4.19	Servicing												●						
4.20	Statistical techniques																		

● Identified relation   ■ Possible relation   ■ Non identified relation

only one or two of these areas. According to the organization's decision, each of these documents should refer to the requirements pertaining to the area or areas under consideration. If an organization is to require the implementation of the three areas, it should develop an EQS Manual. For implementing only the occupational safety and health management system, this Manual (now called the OSH Manual) should only contain information pertaining to the occupational safety and health issues. The same procedure could be applicable to the other documents of the management system. This structure is 'open' and can be used in different ways according to the needs of the organization.

The EQS management manual is the main document of an integrated system. It defines the general rules concerning the environment, quality and occupational safety for the entire organization. It must contain, among other information, the integrated EQS policy, the structure of the organization and the responsibilities of all personnel of the organization who have an influence on E, Q or S. The organization of this manual may follow all the requirements of the proposed system, i.e., it must 'answer' to all the elements of the system. For instance, in the case of quality or the occupational safety and health system, it should 'answer' the twenty elements of the standard.

The EQS general procedures must complement the information of the manual. They must be detailed with the necessary information needed for the most relevant elements of the manual. These procedures must be organised by groups/sections, including: environmental management procedures, quality procedures and occupational safety procedures. A

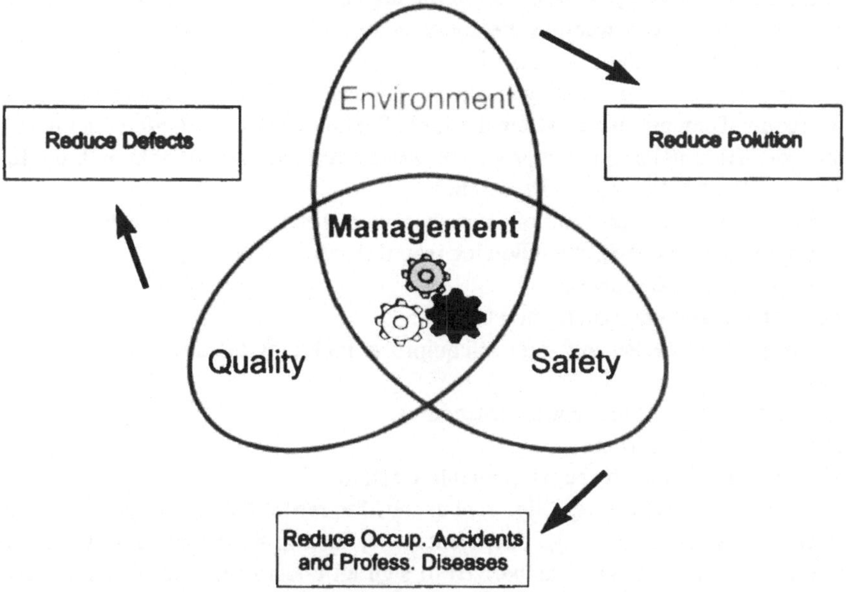

Figure 4. Integration of EQS management systems.

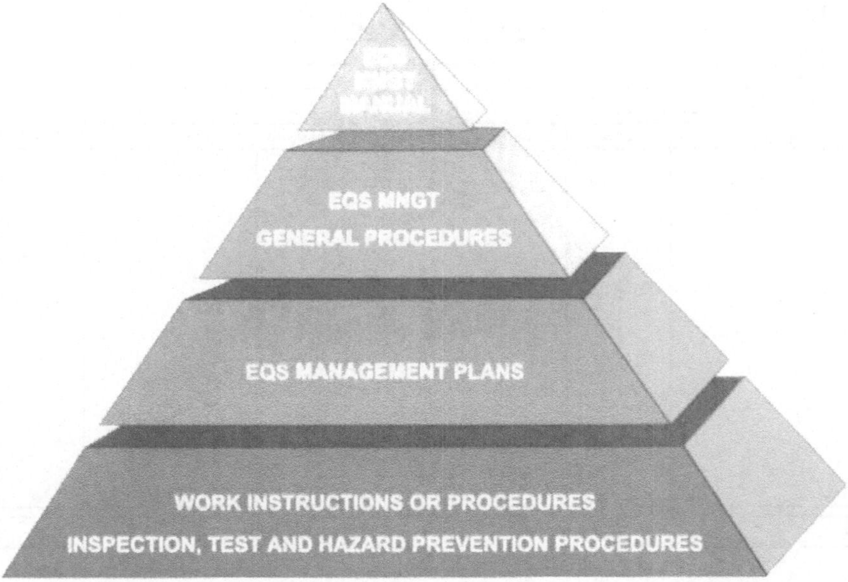

Figure 5. Integration of EQS management system.

formal procedure may include: objective; records of revisions; description of the procedure; responsibilities; reference documents; records generated; and appendices, when applicable.

The EQS management plans are documents with information concerning each particular construction site of the organization. There will be as many plans as there are construction sites within the organization. Like the EQS Manual, each EQS Plan should contain the integrated EQS Policy, the structure of the organization and the responsibilities of all personnel related to the construction site under study.

The EQS management system will also include the Work Instructions or Work Procedures. These are documents describing the technological process concerning each construction operation. They are essential for the identification of hazards (safety and health and environmental hazards) and the appropriate preventive measures to include in the Inspection, Test and Hazard Prevention Procedures.

On the practical level, to implement of the management system on a construction site according to the approach described, the following record sheets may be used:

- Control of signatures and rubrics;
- Control of material and equipment on reception;
- Control of inspection, measuring and test equipment including the auxiliary equipment (IMTE and Auxiliary);
- Inspection, test and hazard prevention procedures;
- Inspection, test and prevention record;
- Nonconformity record and corrective/preventive actions.

The control of signatures and rubrics (Fig. 6) is intended to record the actions of all staff on the construction site authorized to sign documents as defined in the Procedures. The same sheet may be used for all personnel authorized to sign all documents and different sheets

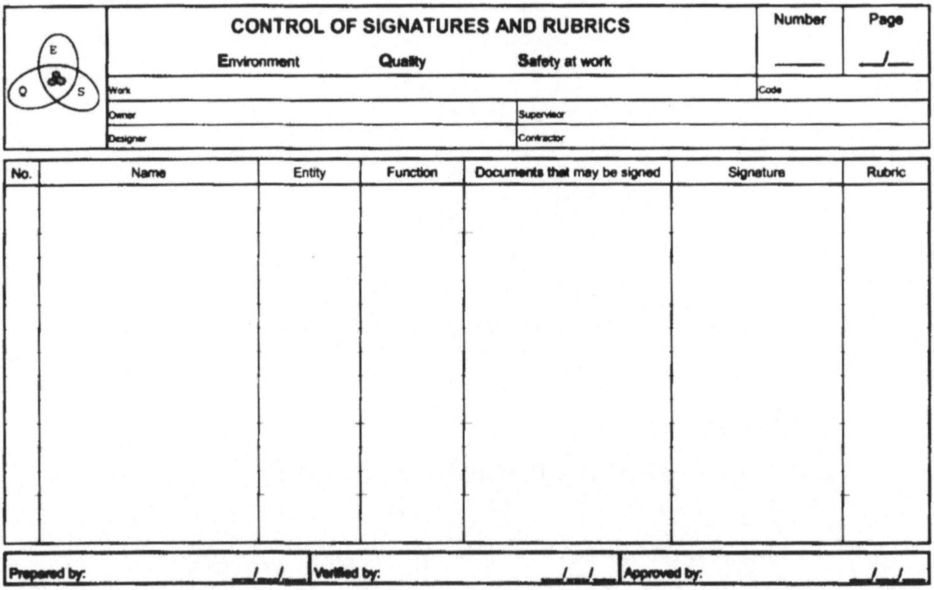

CONTROL OF SIGNATURES AND RUBRICS

Environment Quality Safety at work

Number ____ Page __/__

Work | Code

Owner | Supervisor

Designer | Contractor

No.	Name	Entity	Function	Documents that may be signed	Signature	Rubric

Prepared by: __/__/__ Verified by: __/__/__ Approved by: __/__/__

Figure 6. Control of signatures and rubrics.

CONTROL OF MATERIAL AND EQUIPMENT ON RECEIPT Environment Quality Safety at work		Number ____	Page __/__
Work		Code	
Owner	Supervisor		
Designer	Contractor		

Material or Equipment	Code

Ref.	Checks / Tasks	Objectives / Hazards	Reference documents	Method of checking	Corrective and preventive actions	Resp.

Prepared by: __/__/__	Verified by: __/__/__	Approved by: __/__/__

Reception records					
1 Delivery note n.° ____	2 Delivery note n.° ____	3 Delivery note n.° ____	4 Delivery note n.° ____	5 Delivery note n.° ____	6 Delivery note n.° ____
Received by / Verified by	Received by / Verified by	Received by / Verified by	Received by / Verified by	Received by / Verified by	Received by / Verified by

Figure 7. Control of material and equipment on receipt.

for personnel authorized to sign documents related to only one of the areas under control (environment, quality or occupational safety).

The control of material and equipment on receipt (Fig. 7) must be conducted for all materials and equipment that may be hazardous during their manipulation and / or storage, and require special preventive measures. These hazards, if not addressed, might adversely affect the workers and/or the environment.

The control of inspection, measuring and test equipment including the auxiliary equipment (IMTE and Auxiliary – Fig. 8) must be used to maintain an inventory record of all the equipment existing on the construction site. It must be confirmed that any calibration, guaging or revision is updated. This confirmation is essential for quality and/or for safety. Any equipment not updated must be clearly identified and promptly removed from service.

The inspection, test and hazard prevention procedures (Fig. 9), are intended to be used for all construction operations that may involve hazards. These must be identified as well as the appropriate corrective/preventive actions to be taken. In this sheet the column 'SP' means 'Stop Point'. When postsigned, due to a specific and important check, it means that the work must stop until someone has authorized the procedure and verified the safety conditions of the work. A situation where this can be applied is, for instance, before placing the concrete in a slab form.

The inspection, test and prevention record (Fig. 10) is intended to record the results of all checks that should be performed according to the previously described sheet (the 'inspection, test and hazard prevention procedures' sheet). On a construction project with many piles of the same type there will be, in general, only one 'inspection, test and prevention procedures' sheet. However, for each of the piles a new sheet of 'inspection, test and prevention record' must be used.

CONTROL OF EQUIPMENT (IMTE and Auxiliary)		Number	Page
Environment Quality Safety at work		____	__/__
Work		Code	
Owner	Supervisor		
Designer	Contractor		

REC N°	EQUIPMENT			CALIBRATION / GUAGING / REVISION					GENERAL INSPECTION OF THE EQUIPMENT			Rubric of Resp. for control	Subst. Rec. N°
	Code	Designation	Serial N°	C Q or R	Date of last	On day ? Yes	No	Date of next	Conf	Non conformances detected	Correct until		
					__/__/__			__/__/__			__/__/__		

Prepared by: __/__/__	Verified by: __/__/__	Approved by: __/__/__

Figure 8. Control of equipment (IMTE and Auxiliary).

INSPECTION, TEST AND HAZARD PREVENTION PROCEDURES		Number	Page
Environment Quality Safety at work		____	__/__
Work		Code	
Owner	Supervisor		
Designer	Contractor		

Construction Element / Operation							Code	
Ref.	Checks / Tasks	Objectives / Hazards	Reference documents	Method of checking	Corrective / preventive actions	Resp.	Inspection frequency	SP

Prepared by: __/__/__	Verified by: __/__/__	Approved by: __/__/__

Figure 9. Inspection, test and hazard prevention procedures.

E Q S	**INSPECTION, TEST AND PREVENTION RECORD** **E**nvironment **Q**uality **Safety at work**		Number _____	Page ___/___
	Work:		Code:	
	Owner:	Supervisor		
	Designer	Contractor		

Construction element / Operation	Code
Localisation / Activity:	

Ref. Checks / Tasks	SP	Results	
		Controlled by: Sign	**Verified by:** Sign / /
		Controlled by: Ass /	Verified by. Ass /
		Controlled by: Ass	**Verified by:** Ass
		Controlled by: Ass	**Verified by:** Ass /
		Controlled by: Ass	**Verified by:** Ass
		Controlled by: Ass	**Verified by:** Ass
		Controlled by: Ass	**Verified by:** Ass
		Controlled by: Ass	**Verified by:** Ass :

Mod. AQ805

Figure 10. Inspection, test and prevention record.

When performing the checks mentioned in the above sheets, there will be situations of nonconformity (for instance, due to deficiencies detected on a safety net or other protection; non application of foreseen measures, etc.). In these cases corrective and / or preventive actions should be taken and they must be recorded in sheets similar to the one presented in Figure 11.

5 EVALUATION OF (NON) SAFETY AND HEALTH AND ENVIRONMENT ON CONSTRUCTION SITES

The implementation of safety and health and environmental management systems on construction sites may be evaluated through audits. They are intended to check the conformity of the system's activities.

These audits are generally conducted based on the international standards ISO 10011 (quality) or ISO 14011 (environment). These standards are similar and it is expected they will be fully compatible in the near future (see comments in the section 'Management System Standards').

To conduct an audit the following three documents must be prepared:

- Audit plan;
- Audit record;
- Audit report.

The Audit Plan must be prepared by the audit team and be approved by the client and communicated to the auditee. It must include:

- Objectives and scope;
- Identification of the individuals within the auditee's organization having significant direct responsibilities regarding the units to be audited;
- Identification of reference documents (standards, quality manual, quality plan, safety and health plan, etc.);
- Identification of audit team members;
- The working and reporting languages of the audit;
- The date and place where the audit is to be conducted;
- Identification of the units to be audited;
- The schedule of meetings to be held with the auditee's management;
- Confidentiality requirements;
- Distribution of the audit report and expected date of issue.

The Audit Record must also be prepared by the audit team and should contain a check list of the issues to be analysed during the auditing process. An example of this audit record is presented on Figure 12.

The Audit Report must be prepared under the direction of the Lead Auditor, who should sign and date it, and send it to the client. It must include:

- Introduction
  Scope; objectives; audit units; audit team members; reference documents, etc..
- Audit development
  Description of all developed actions and analysis conducted, etc..
- Conclusions
  Conclusions of each item of the audit; identification of nonconformances, obstacles encountered, etc..

E Q S	**NONCONFORMITY RECORD AND CORRECTIVE / PREVENTIVE ACTIONS** **Environment** **Quality** **Safety at work**		Number _____	Page ___/___
	Work:		Code	
	Owner:	Supervisor:		
	Designer:	Contractor:		

Description of nonconformity:

Localization:

Reference documents:

Described by: ___/___/___ Verified by: ___/___/___

Description of corrective / preventive actions:

Correction until:

Proposed by: ___/___/___ Approved by: ___/___/___

Execution of corrective / preventive actions:

Executed by: ___/___/___ Controlled by: ___/___/___

Verified by: ___/___/___ Approved by: ___/___/___

Mod. AQ808

Figure11. Nonconformity record and corrective/preventive actions.

	AUDIT RECORD	Página: 1/

INTERNAL AUDIT ☐	***EXTERNAL AUDIT*** ☐
Audit scope:	

Entity:	
Unit	
Responsible:	
Place of the audit:	

Audit team
1.
2.
3.

Start	Day	Time	*End*	Day	Time:

REF.	Description	CONFORM ?			Remarks
		Yes	Part	No	

Figure 12. Example of an audit record sheet.

Table 6. Accident and accident indices record.

Date		Average n° of workers		Man × hours worked		N° of accidents				N° of days lost		Incidence Index		Frequency Index		Severity Index		Duration Index	
						Fatal		Non-fatal											
Year	Month	Month	Total	Month	Total	Month	Total	Month	Total	Month	Total	Month	Total	Month	Total	Month	Total	Month	Total

Incidence Index	Frequency Index	Severity Index	Duration Index
$II = \frac{\text{No. of Sccidents} \times 1000}{\text{No. of Workers (average)}}$	$FI = \frac{\text{No. of Sccidents} \times 1000000}{\text{No. of hours worked}}$	$SI = \frac{\text{No. of days lost} \times 1000}{\text{No. of hours worked}}$	$DI = \frac{\text{No. of day lost}}{\text{No. of lost day accidents}}$

- Actions to be developed
  Description of actions due to the conformances encountered, etc...
- Annexes
  Records of attendees at the meetings and other records.

Concerning the evaluation of safety and health, one can use the Accident and Accident Indices Record presented in Table 6. This information may be recorded and calculated for a single construction site, for a group of construction sites (within a department), and for all construction sites of an organization. It may also be used to record the performance within a country. The use of standard indices allows comparisons between different entities.

6 CONCLUSIONS

In this chapter the possible integration of environmental, quality, and occupational safety and health management systems was presented and discussed, focusing mainly on the areas of environment and occupational safety and health on construction sites. It summarizes a study under development at the Department of Civil Engineering of Instituto Superior Técnico – Technical University of Lisbon.

The study attempts to contribute to the discussion on the need for a single methodology to implement the three areas under consideration on the construction sites. It is based on the ISO 9000 requirements taking into account the specific environmental and occupational safety and health concerns. Although more work needs to be done on this subject, it is the author's belief that the benefits that can be achieved with such integration justify open discussion.

Recent decisions of ISO on these matters and the fact that some countries or institutions are developing guides for the implementation of occupational safety management systems, strengthen the need for this discussion.

It should be pointed out that some of these countries or institutions are developing these guides based on the ISO 14000 family of standards and others are basing the development on the ISO 9000 family of standards. These different ways of developing these standards does not favor the international market, and may create difficulties between countries that have relationships, such as is the case of the European Union.

If quality is recognized as an essential element of the competitive climate, the same is also true of occupational safety and health and the environment. Preserving the lives of construction workers and the environment is an objective that should be aggressively pursued as we enter the next millennium.

REFERENCES

Levitt, R. E. & Samelson, N. M. 1993. *Construction Safety Management*. John Wiley and Son, New York, USA.

Eubanks, F. M. 1994. Elements of a Construction Safety Management Program. *Fifth Annual Rinker Internacional Conference on Building Construction focusing on Construction Safety and Loss Control*. University of Florida. USA.

International Organization for Standardization (ISO). Family of Quality Management Standards (ISO 9000).

International Organization for Standardization (ISO). Series of Environmental Management Standards (ISO 14000).

Alves Dias, L. M. & Fonseca, M. 1996. *Safety and Health Plan in Construction* (Portuguese version). IST-IDICT. Lisbon, Portugal.

Cascio, J. et al. 1996. ISO 14000 Guide – The new International Environmental Management Standards. McGraw-Hill. USA.

American National Standards Institute – ANSI 1996. News release (www.ansi.org).

British Standards Institution 1996. BS 8800 – Safety Management. Guide to Management Principles. London, UK

Standards Australia/Standards New Zealand 1996. Occupational Health and Safety Management Systems – General guidelines on principles, systems and supporting techniques. Sydney, Australia.

American Industrial Hygiene Association 1996. Occupational Health and Safety Management System: An AIHA guidance document. Virginia. USA.

American Industrial Hygiene Association – AIHA 1996. New Frontiers in Occupational Health and Safety Management System: A Management Systems Approach and the ISO Model. Virginia. USA.

Hinze, Jimmie W. 1997. *Construction Safety*. Prentice Hall. New Jersey, USA.

International Organization for Standardization – ISO (1997, 1998). Press release (www.iso.ch).

Marcus, P. et al. 1997. Moving ahead with ISO 14000 – Improving Environmental Management and Advancing Sustainable Development. Wiley & Sons. USA.

Alves Dias, L. M. & Pires, J. 1998. Quality and Occupational Safety in Construction (Portuguese version). IDICT, Lisbon, Portugal.

CHAPTER 3

The health and safety protection plan and the file containing features of the building according to EEC Directive 92/57

JOACHIM BERGER
LL.D, Germany

ABSTRACT: In none of the 12 member states of the European Community (EEC) is it possible to sit back and claim that safety and health protection on construction sites has reached the stage where it could be accepted as a calculable risk. About 7,5% of all gainfully employed persons in the EEC work in the construction industry, accounting for 15% of all accidents at work, and 30% of all fatal industrial accidents. Has this situation arisen because those who work in the construction industry are less safety conscious than employees employed in other sectors? The construction industry, differs substantially from other branches of the economy in several respects. In this chapter some of these differences are discussed. The impact of Council Directive 92/57/EEC which was passed in 1992 is evaluated as a response to the poor safety and health record of the European construction sector. This directive was issued as part of the commitment of the Council of European Communities to the adoption of minimum requirements for encouraging improvements in working environments to ensure a better level of protection of the safety and health of workers. All European Union member states were to implement the terms of the directive into national legislation by 1994. In Germany the ordinance only came into force during 1998, despite Great Britain, France, and the Netherlands having done so several years ago.

1 INTRODUCTION

In none of the 12 member states of the European Community (EEC) is it possible to sit back and claim that safety and health protection on construction sites has reached the stage where it could be accepted as a calculable risk. About 7,5% of all gainfully employed persons in the EEC work in the construction industry, accounting for 15% of all accidents at work, and 30% of all fatal industrial accidents. Against this background, the construction industry can arguably be considered one of the high-risk economic sectors.

Why is this the case?
Is it because those who work in the construction industry are less conscious than employees employed in other sectors?

Certainly not!

The construction industry, however, differs substantially from other branches of the economy in several respects:

- Every construction site is subject to local conditions, and is unique in form, size and purpose.
- Each construction site represents in effect the creation of a production site where new workplaces are set up. It may be possible to speak of mobile factories.
- The newly set up workplaces are subject to construction change. The workers are required to familiarise themselves almost on an hourly basis with new situations. This change is generated to some extent due to the construction progress of the building project, but also due to the interaction over time and space of several firms on the construction site.
- Additionally, particularly in the central and northern reaches of the EEC, there is a need to adapt to constantly changing weather conditions.
- The construction industry is primarily made up of a large number of small companies and self-employed workers (one-man businesses), which are often neither in the position to be aware of the complexity of building operations nor able to recognise the hazards which could arise from them.

2 POSSIBLE SOLUTIONS TO SAFETY AND HEALTH PROBLEMS ON CONSTRUCTION SITES

Systematic steps in hazard control

- Analyse the hazards;
- Define the protection targets;
- Lay down precautions, which can meet the protection targets;
- Implement the precautions; and
- Control the effects of precaution on safety and health at work.

Hierarchy and priority of protection targets

- Eliminate the hazards;
- Avoid all the effects of existing hazards, for example, by separation and covering against hazards and removing the person from the danger area;
- Avoid and reduce the effects of hazards on the workers, for example, by protection of workers by using personal protective equipment; and
- Manipulate the human behaviour at the workplace.

In summary, it is not for those who are affected by the hazards to select the method of protection against these hazards. The first level in the hierarchy should be attempted first and be carefully examined and evaluated, and only then, when this level cannot be completely realised, should the next level be attempted.

While this is familiar territory, it does not, *prima facie,* offer any starting point for improvement to safety and health on construction sites.

3 A QUESTION OF PRINCIPLE

The current discussion in economic circles in recent times is around the concepts of 'lean management' and 'lean production.' This is evidently the direction being taken by other economic sectors.

Is this approach also being followed in the construction sector?
The larger firms are certainly implementing both concepts. However it is not being adopted by the many small firms with less than 20 workers, which make up 95% of the firms operating in the construction industry.

What has this got to do with safer workplaces? Nothing really!
It is a question of the principle involved. It is a question of giving consideration to and solving problems at a very early stage. Throughout the economy the preparations for a job of work are a pre-condition for its success. The construction industry is no different, particularly with respect to working methods and timetables or programme.

However, when working processes are examined from the point of view of safety and health on the construction site, decisions on many safety and health issues are left to the eleventh hour. The implication is that those responsible for safety on sites are presented with situations, which are very complicated – sometimes even unsolvable.

There are regrets every time that preparations for the work have not covered workplace safety – yet every time the same mistake is repeated when the next building project is planned.

What about the horrendous costs involved? Every anticipated solution, even the simplest, which is to be effected on construction sites due to an earlier oversight, is considerably more expensive than any solution which could have been planned and carefully evaluated.

Having taken cognisance of this, while at the same time resolving that the same mistakes will not be repeated, two questions arise:

1. When should matters of safety and health be addressed in any building project?
2. How is this accomplished?

The answers to both are simple.

The only response to the first question is that matters of safety and health must be addressed as soon as possible – at the project planning stage. The answer to the second question is embodied in the motivation for, and the requirements of the Council Directive 92/57/EEC. The health and safety plan and the health and safety file are 2 vehicles to achieve improved safety and health performance on construction sites.

4 BACKGROUND TO COUNCIL DIRECTIVE 92/57/EEC

The Council Directive 92/57/EEC was passed in 1992. This directive was issued as part of the commitment of the Council of European Communities to the adoption of minimum requirements for encouraging improvements in working environments to ensure a better level of protection of the safety and health of workers. All European Union members were to implememnt the terms of the directive into national legislation by 1994.

Resistance to change in any form is normal and to be expected. Reaction to this directive was no different. The conversion into national law in many European countries, Germany included, took some time because architects, in particular, across Europe felt very uncomfortable with this change in responsibility from the contractor to the client who was required to take appropriate steps with respect to safety and health in the planning and execution of a construction project. In Germany the ordinance only came into force during 1998, despite Great Britain, France, and the Netherlands having done so several years ago.

Further, concern was expressed about every construction site would now requiring a formal health and safety plan which could conceivably extend to between 70 and 90 pages, for example. It was argued that the health and safety plan would place an unreasonable demand on funds and time with no practical value on most construction sites.

Motivation for the directive stemmed from the recognition by the Commission that more than 50% of occupational accidents on construction sites had been attributed to unsatisfactory architectural and/or organisational options or poor planning of the works at the project preparation stage. Moreover, the Commission recognised that large numbers of accidents may be caused by inadequate co-ordination especially where various undertakings worked simultaneously or in succession at the same construction site.

The enactment of regulations is the first step. Enforcement in the construction industry is the next.

5 PRACTICAL IMPLICATIONS OF COUNCIL DIRECTIVE 92/57/EEC

What does the Directive cover and what changes does it bring about?
In terms of the Directive, the client must take steps in the planning and execution of a building project and is responsible for organising the work on the construction site in such a way that risks to life and health were avoided as far as is possible, and where not possible, to maintain residual risk at the lowest level. The responsibility for taking these measures may be entrusted to a third party.

5.1 *Safety and health co-ordinators*

The directive requires one or more safety and health co-ordinators to be appointed by the client if workers from multiple employers will be working on the same construction site. Both a safety and health plan as well as a safety and health file are to be drawn up by the safety and health co-ordinator. The functions of the co-ordinator may be performed by the client if this is preferable and desirable. These co-ordinators, where they are appointed, through the execution of their duties, do not relieve the client of any responsibilities in terms of the construction project.

5.2 *Preliminary notification*

Preliminary notification must be given to the responsible factory inspectorate at least 2 weeks before work is due to commence on the construction site. This preliminary notification is required for safety and health at work on all construction sites where the work is scheduled to last longer than 30 working days and on which more than 20 workers are employed at the same time, or on which the amount of work to be carried out is scheduled to be more than 500 person-days. This notification must be visibly displayed on the construction site.

Some of the information to be submitted in this notification includes:

- The exact physical address of the site;
- The details of the client(s);
- The type of project;
- The details of the safety and health co-ordinators;
- The commencement date for work on site;
- The planned duration of the project;
- The estimated maximum number of workers who will work on the site; and
- The planned number of contractors and self-employed persons who will be working on the site.

5.3 *Health and safety plan*

Additionally, the client is responsible for the compilation of a health and safety plan during the project planning stage before actual work begins on site. The safety plan must take into account all the working safety provisions to be applied during the erection of the building. The health and safety plan is necessary

- If workers from multiple employers will be working on the same construction site and a preliminary notice is needed; or
- If workers from multiple employers will be working on the same construction site and particularly dangerous work is to be executed such as work in which the employees are exposed to a danger of falling more than 7m.

Additionally, the safety plan must name hazards that are specific to the construction site concerned, list regulations which contain solutions for avoidance of those dangers and must make concrete suggestions of solutions to avert hazards. The safety plan must take into account the work involving particular risks listed in Annex II of the directive.

Further, the plan should be drawn up according to trades or trade groups. The consequences of the solutions found must be expressed in particular items in the invitation to tender. In this way the safety plan becomes an integral part of the tender or bidding process and becomes an element of the contract itself.

Advantages of this procedure include:

- Risks are recognised and identified before the work begins, eliminating any potential excuse of ignorance of these risks; and
- Solutions which might avert the risk are suggested and find their expression in specific items in bills of quantities, for example, eliminating the posibility for contractors to claim that the equipment or procedures are too costly since they would have priced for them as part of their tender or bid proposal.

The implications. Further, are that dangerous situations, timetable or programme delays, and unforeseen costs are avoided and costs reduced. Savings were achieved by following this process on the EEC model construction site in Nürmberg, Germany. At this project where the budget was about 50 million DM ($30 million) it took about an hour to complete the health and safety plan for each trade.

5.4 *Health and safety file*

The safety and health co-ordinators are required to compile a health and safety file to be handed to the client upon completion of the project for use in the course of later maintenance and repair work.

The following information should be included in the health and safety file:

- Historic site data;
- Site survey information;
- Site investigation reports and records;
- Photographic record of essential site elements;
- Statement of design philosophy, calculations and applicable design standards;
- Drawings and plans used throughout the construction process, including drawings prepared for tender purposes;
- Record drawings and plans of the completed structure;
- Maintenance instructions;
- Instructions on the handling and/or operation of equipment together with the relevant maintenance manuals;
- Results of proofing or load tests;
- Commissioning test results;
- Materials used in the structure identifying, in particular, hazardous materials including data sheets prepared and supplied by suppliers;
- Identification and specification of in-built safety features, for example, emergency and fire fighting systems and fail safe devices; and
- Method statements produced by the principal contractor and/or contractors.

5.5 *Obligations of employers*

The directive in no way absolves employers from their responsibilities towards their workers.

6 THE PROCEDURE TO BE FOLLOWED FOR THE HEALTH AND SAFETY PLAN

The suggested pattern for the compilation of the health and safety plan simplifies the procedure considerably (Table 1). This pattern should allay the fears that these plans would be voluminous documents, be time-consuming to prepare, costly and have no practical value.

The particular trade is identified and inserted at the top of the page. Similarly, the particular building project under consideration is identified.

The risks which could occur relative to a particular trade or group of trades are listed in column 1. For earthwork, for example, the question would be considered of whether overhead lines and buildings are present in close proximity to the proposed building works.

Column 2 is used to identify at which point in time the solutions to this risk must be decided upon and is to be recognised as an aid to the co-ordinator. Further, if the hazards applies to the particular building project an 'x' is inserted in column 3.

In columns 7 and 8 the type and title of the applicable regulations to be observed with respect to the hazard are entered.

In the other columns the possible and actual solutions to the hazard are, or will be, entered by means of an 'x.' In column 9 all the suggested solutions arising from the regulations in column 8 are listed. The co-ordinator, in consultation with the designers (architects and/or building engineers), selects the optimal solution for dealing with the hazard.

Table 1. An example of the suggested pattern for health and safety plans.

Trade		Brickwork				Building site					
		Is the hazard source present?		Hazard to health and safety		Regulations to be considered		Solutions			
Source of hazard	Decision date	Yes	No	Of employee	To third party	Type	Title	Possible	Selected	Changed	Remarks
1	2	3	4	5	6	7	8	9	10	11	12

In this way the designers are asked the appropriate question at the appropriate time – as early as possible. The question is addressed in respect of the particular construction site, for a specific trade, and for a particular hazard. A solution is therefore timeously found which will avert the hazard. This solution is marked off in column 10.

The text for the invitation to tender is defined in column 12. If, during the actual project execution, a better or more cost-effective solution is identified, the co-ordinator after consultation with the designers and the contractor, will indicate this change in column 11 and will define the newly selected performance in column 12.

While at first glance this procedure seems very complicated, it is really very simple.

7 EXPECTED BENEFITS OF THE DIRECTIVE

The building site directive will clearly improve safety at work on construction sites. It has been estimated that accidents on smaller construction sites will be reduced by about 33,3% and on larger construction sites by 20%. The changes which the directive promotes could result in an increase in profitability in the region of 15 billion DM ($9 billion).

REFERENCES

Council Directive 92/57/EEC 1992. Council Directive 92/57/EEC of 24 June 1992 on the implementation of minimum safety and health requirements at temporary or mobile construction sites (eighth individual Directive within the meaning of Article 16 (1) of Directive 89/391/EEC) Official Journal of the European Communities no. L 245/6.

CHAPTER 4

Analysis and control of human errors during structural construction

XILA LIU
Department of Civil Engineering, Tsinghua University, Beijing, P.R. China

ABSTRACT: A simulation model of structural construction including the checking process is presented. The main parameters of reinforced concrete structures influenced by construction human errors can be simulated in this model. A structural reliability analysis method is also proposed in this paper, which takes the combined human errors into account. Using the present method, the efficiency of different control measures to control the human errors is analyzed. A satisfied control scheme to achieve an reasonable failure probability can be decided through the analyses and proposed program. Sensitivity of structural resistance to human errors that influences the structural parameters and the occurring possibility of human errors is used to choose the key control points of human errors. Finally, an example is shown.

1 INTRODUCTION

Human errors occurred during plan, design and construction of structures are the main causes of structural failures. In this case, they should be taken into account in structural reliability analysis. In China, human errors occurred during construction are especially serious, at which people are extremely concerned. The main purpose of this chapter is to analyze the influence of construction human errors to structural reliability and to control the errors more efficiently.

To control the destructive effects of human errors efficiently, some problems must be solved first, such as: Where are the key positions of a structure to be controlled during construction? Which one is the key parameter to be controlled? How to control the key points efficiently? The architecture adopted in this study is shown as follows (Fig. 1).

2 SIMULATION OF HUMAN ERRORS INFLUENCE TO PARAMETERS

Since human errors occurred during construction are the main causes of structural failures, it is necessary to establish human error models for structural reliability analysis and human error control. Human Reliability Analysis (HRA) is considered as a suitable method for this purpose. This method is widely used in safety analyses of nuclear power plants, and now it

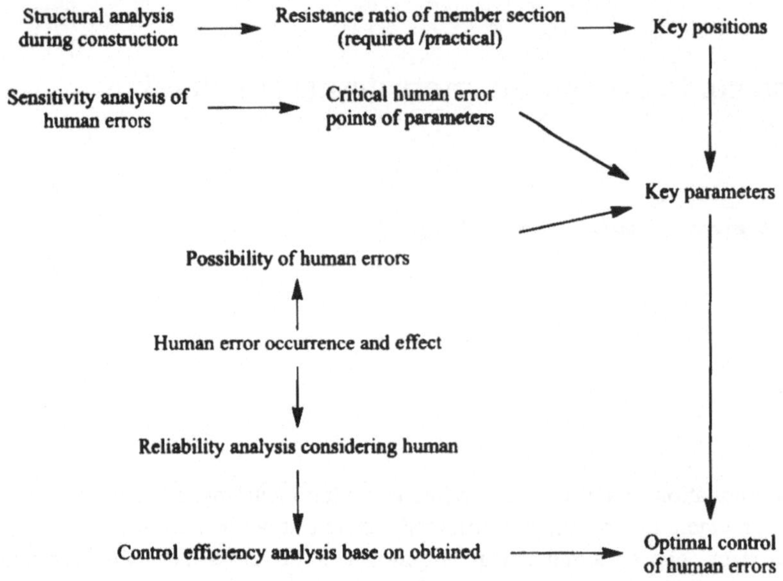

Figure 1. Study architecture.

has been used in structural engineering (Nessim 1983, Nowak 1985). Though it is very difficult to measure human errors accurately (because of the complexity of human behavior) we can evaluate the relative effects of the different human errors using the HRA method.

2.1 *Expression construction operation*

Any construction process can be divided to a series of simple operations. The event tree of a basic operation is shown in Figure 2.

An operation has two results, no error (NE) or human error occurring (HE). Different from the simple two alternation (false or true) in device operation, more complex choice must be concerned in construction operation. When a human error occurred, it may have a changeable influence to structural parameters, and its influence may be conservative (CE) or dangerous (DE) to structural safety. If the occurring probability of a human error and its influence extent to the relative parameters is known, the operation can be simulated in a computer.

The influence process of a construction operation is shown in Figure 3.

2.2 *The occurring probability model of human errors*

The occurring probability of a human error is expressed as human error rate P_E: $P_E = N_E/N$, where N is the number of operations, N_E is the number of error operations. The human error rate is a random variable. A skilled operator may make less error. The logarithmic normal distribution is proposed to describe the human error rate during construction by Swain etc. The model is also widely used to express human error rate in safety analyses of nuclear power plants. The probability density function of human error rate is:

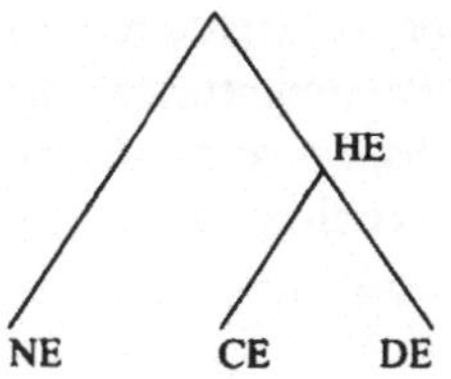

Figure 2. Event tree of a operation.

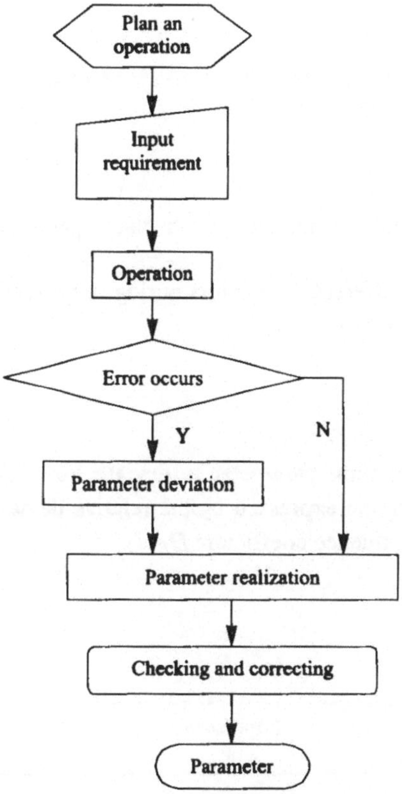

Figure 3. Influence process of operation error.

$$f(P_E) = \frac{1}{\xi \cdot P_E \sqrt{2\pi}} e^{-\frac{(\ln P_E - \lambda)}{2\xi^2}} \quad (1)$$

$$\xi^2 = \ln(1 + \sigma^2/\mu^2) \quad (2)$$

$$\lambda = \ln \frac{\mu}{\sqrt{1 + \sigma^2/\mu^2}} = \ln(P_{Em}) \quad (3)$$

where μ, P_{Em}, σ and ξ are the expected value, mean value, the standard deviation of human error rate and the corresponded disperse parameter respectively. They are the parameters that represent average human error occurring rate and dispersive degree respectively. They reflect the effect of operator's ability, personality and work circumstance. The average error rate can be obtained from investigation. To have the dispersive parameter, much more work must be done. A simple parameter describing the uncertainty of the error rate, which is defined as error factor EF (Apostolakis 1982), is adopted here:

$$EF = \sqrt{\frac{\Pr(F_{90th})}{\Pr(F_{10th})}} \tag{4}$$

where $\Pr(F_{90th})$ and $\Pr(F_{10th})$ are the probability of human error rates corresponding to the 90% and 10% respectively. Then ξ is (Apostolakis 1982):

$$\xi = \ln(EF)/1.2817 \tag{5}$$

Swain etc. proposed a general rule of human error rate for nuclear power plant operators in ordinary condition, which is shown in Table 1.

The occurrence rate of human errors influences different parameters during reinforced concrete structure construction, which is shown in Table 2.

2.3 *Human error influence extent model*

When a human error occurred, it may cause a structural parameter to deviate from its normal value. The human error influence extent may be expressed by the relative deviation of parameter, which is defined as human error influence coefficient D_E:

Table 1. Error factors.

Estimate of error rate HEP	Error factor EF
< 0.001	10
0.001–0.01	5
> 0.01	3

Table 2. Occurrence rate of human error.

Error code	Error type	Average error rate HEP	Error factor EF	Average error rate HEP_I (after checking)	Error factor EF_I (after checking)
E1 (Park 1987)	Misreading error	0.003	3		
E2 (Stewart 1993)	Increase of section height	0.0188	5	0.000321	12.08
E5 (Stewart 1993)	Decrease of section height	0.0296	5	0.000506	12.08
E6 (Stewart 1993)	Increase of section width	0.0083	3	0.000142	9.00
E7 (Stewart 1993)	Decrease of section height	0.0081	3	0.000139	9.00
E8 (Xu Maobo 1996)	Insufficiency of concrete strength	0.22	3		

$$D_E = \frac{X_E - X_0}{X_0} \tag{6}$$

where X_0 is the normal value of parameter without human error, and X_E is the error-affected value, which are random variables. There is no general accepted model about the human error influence extent to structural parameter. Because D_E is associated with multiple factors, a logarithmic normal distribution is supposed.

The distribution parameters of human error influence coefficient D_E is shown in Table 3.

2.4 *Human error checking*

The Signoid response function, which is widely used in artificial neural networks, is introduced as the human error checking function. It means that larger errors will be more easily checked out, the increase of discovery probability along with the influence extent of error is an *S* shape curve (Fig. 4). For the serious errors that influence extent exceed a threshold, the discovery probability is a constant.

$$P(D_E) = \begin{cases} \dfrac{1}{1+e^{(\alpha-\lambda D_E)}} & D_E \le D_{EL} \\ P_L & D_E > D_{EL} \end{cases} \tag{7}$$

where, D_E is the measure of human error influence to structural parameters; P is the discovery probability of human error; λ and α are the constants reflect the characteristic of $P(D_E)$; D_{EL} is the dividing point of serious errors, P_L is the discovery probability of serious errors.

2.5 *Simulation*

The flow chart of simulating the influence of construction human errors to a structural parameter is shown as Figure 5.

The checking simulation flow chart is as Figure 6.

Based on above model, the influence of human errors to structural parameters can be simulated in the computer. Different from those models proposed earlier, the simulation model given herein considers both the randomness of human error rate and influence extent of the human error simultaneously. In the simulation, Monte-Carlo method is used.

Table 3. Distribution characteristic of D_E.

Error code	Error type	95% believe lower limit	average	95% believe higher limit
E2 (Stewart 1993)	Increase of section height	4.8	6.27	8.19
E3 (Stewart 1993)	Decrease of section height	–5.44	–7.10	–9.28
E4 (Stewart 1993)	Increase of section width	3.95	5.22	6.89
E5 (Stewart 1993)	Decrease of section height	–4.03	–5.24	–6.81
E6 (Xu Maobo 1996)	Insufficiency of concrete strength	–7.51	–9.58	–12.22

The comparisons of simulation and practice investigation data of concrete strength are shown in Figure 7 and Table 4 respectively.

All the investigation and simulation results are original, the comparison of simulation results with investigation data seems satisfied.

The simulation results of the cross section height of reinforced concrete beams are listed in Table 5. Different checking states are considered here. According to the requirement of construction quality control, at least two times of checking is needed. Because the checking simulation by computer is much more strict to follow the requirement set up previously, the remained human error rate of simulation results is less then that of the investigation results. But they are at a same level.

3 STRUCTURAL RELIABILITY ANALYSIS INCLUDING HUMAN ERRORS

A structural reliability analysis approach considering human errors is proposed in the present study. The human errors is assumed as binary choice in a particular operation, occurring or not occurring, and the effect magnitude of human error on structural parameter is considered as random variable in the model. Combined errors of different types are included in the analysis.

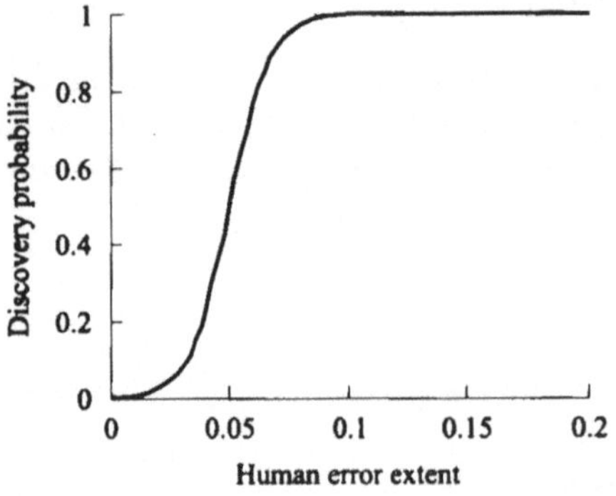

Figure 4. S shape function.

Table 4. Comparison of simulation and investigation data on concrete strength.

Parameter	Investigation	Simulation	Difference(%)
μ	22.77	22.8	0.13
σ	4.34	4.63	5.91

Table 5. Simulation and investigation results on section checking.

State of checking	Misreading error rate	Conservative error rate	Damage error rate
Before check	0.003071	0.041076	0.063267
First check	7.00E–06	0.003525	0.005237
Second check	0	0.001863	0.002854
Investigation	No data	0.00218	0.00335

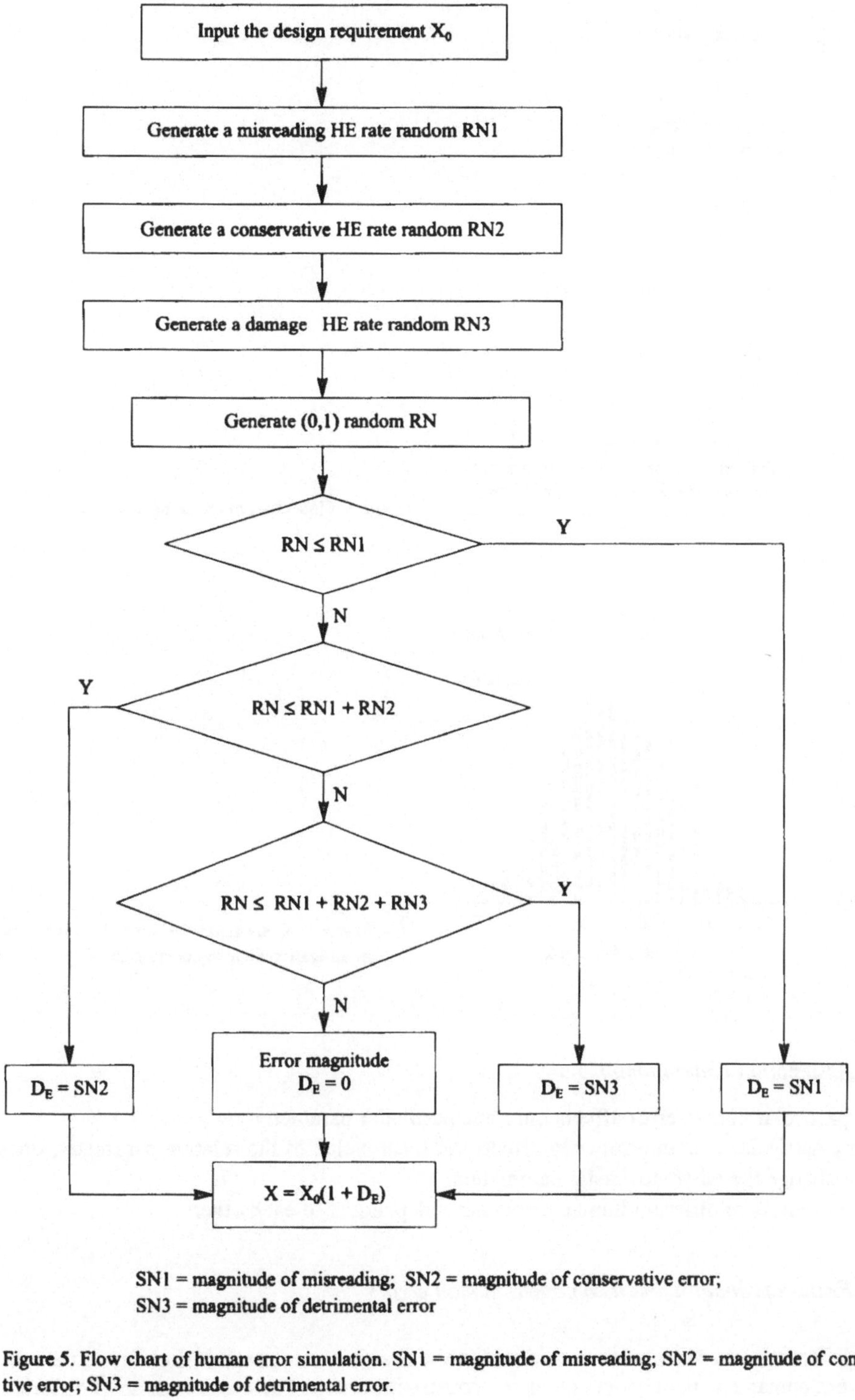

Figure 5. Flow chart of human error simulation. SN1 = magnitude of misreading; SN2 = magnitude of conservative error; SN3 = magnitude of detrimental error.

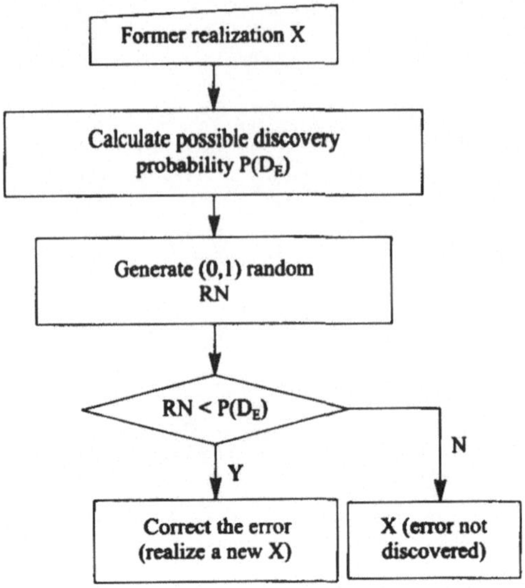

Figure 6. Flow chart of checking simulation.

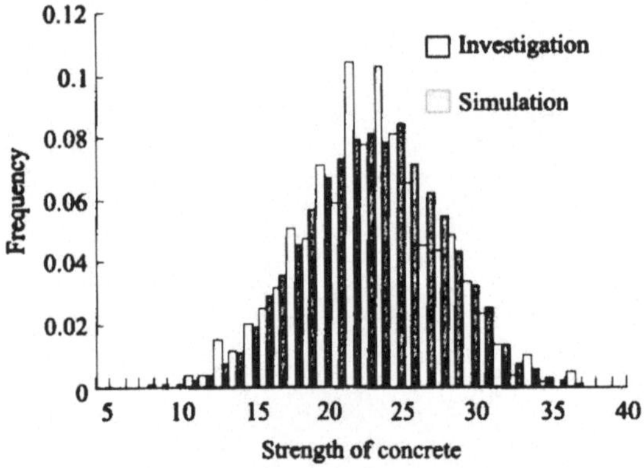

Figure 7. Comparison of simulation data and investigation data on concrete strength.

3.1 *Assumption about human errors*

- A particular human error affects only one particular parameter.
- Any particular human error only affects the mean value of the relative parameter, does not change the other stochastic parameters.
- Occurrences of different human errors are independent of each other.

3.2 *Failure probability with combined human errors*

If m different human errors are considered in the analysis, it is possible that any l $(l \leq m)$ different human errors may occur at same construction step. The event with any l different human errors occurring at same time is described as

$$HE_{cjl} = \bigcap^{l} HE_{ji} \tag{8}$$

where, subscription *cjl* means the combined *l* human error events, *i* refers the numbering of a particular human error, and *j* refers some combination form of different human errors. Because of the independence of different human errors, the probability of event HE_{cjl} can be written as

$$P_{HEcjl} = \prod^{l} P_{HE_{ji}} \tag{9}$$

According to a certain combination *j*, the combined probability density function of the event HE_{cjl} is $\prod^{l} f_{ji}(X_{ji})$, where, X_{ji} is the effect magnitude of the human error *i* with the combination *j* to the relative parameters. For all $l \le m$, there are *n* different possible combinations in total. The general human error event can be described as follows

$$HE = \bigcup^{n} \bigcap^{l} HE_{ji} \tag{10}$$

and the overall probability of human error occurring is

$$P_{HE} = \sum^{n} \prod^{l} P_{HE_{ji}} \tag{11}$$

The error-free event is written as HE_0, and its occurrence probability is

$$P_{HE_0} = 1 - P_{HE} \tag{12}$$

Considering error-free event and all possible combined human error events, the total structural failure probability can be shown as follows

$$P_f = P_{HE_0} \cdot P_{fHE_0} + P_{HE} \cdot P_{fHE} \tag{13}$$

where, P_{fHE_0} is the failure probability of error-free structure, P_{fHE} is the failure probability considering all possible errors and is written as

$$P_{fHE} = \sum^{n} P_{HE_{cjl}} \cdot P_{fHE_{cjl}} \tag{14}$$

where the failure probability of structure $P_{fHE_{cjl}}$ when human error event HE_{cjl} occurs is written as follows

$$P_{fHE_{cjl}} = \underbrace{\int\int\cdots\int}_{l} \prod^{l} f_{ji}(x_{ji}) \cdot P_f(x_{ji}) \cdot dx_{ji} \tag{15}$$

in above equation, $P_f(x_{ji})$ is the failure probability of structure when *l* human errors HE_{ji} occur and influence the parameters with x_{ji}.

Monte-Carlo method is used to calculate the failure probability with a given human error, and a simplified approach is proposed to calculate the above integral equation.

4 INFLUENCE OF HUMAN ERRORS ON STRUCTURAL RELIABILITY AND CONTROL EFFICIENCY

Using mentioned method for calculating structural reliability considering human errors the reliability of a typical bending section of reinforced concrete beams under different measures is analyzed. The control measures include that, such as, checking in different times, different checking sampling rates at a time and increase of some structural parameters. The calculating results are as follows.

4.1 *Results of different checking times*

The calculating results of failure probability under different times of checking are listed in Table 6 and the shown as Figure 8. It is found that the first checking is much more efficient.

The calculating results of failure probability under different sampling rates in first checking are listed in Table 7 and shown in Figure 9.

It is found that the failure probability reduction with the increase of sampling rate is a linear relation. The calculated results of second and third times checking are also similar.

4.2 *Checking efficiency analysis*

The definition of checking efficiency in the present paper is that: the relative decrease of failure probability resulted from a unit increase (say 10%) of the checking sampling rate.

The calculating results of the checking efficiency based on different sampling rates during first and second checking are shown in the Figure 10.

It is found that in the first checking the checking efficiency improves significantly with increasing of the total checking sampling rate, but in the second checking (assuming that the first time checking sampling rate is 100%) the checking efficiency is almost un-

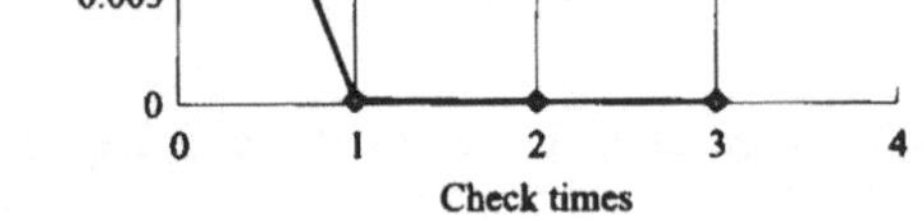

Figure 8. Results of checking times.

Table 6. Results of different checking times.

Checking times	Before checking	Checking once	Checking twice	Checking three time
Failure probability value	0.012266	1.01E–04	6.23E–05	6.22E–05
Related reduction of failure probability %		99.17996	38.06234	0.130979

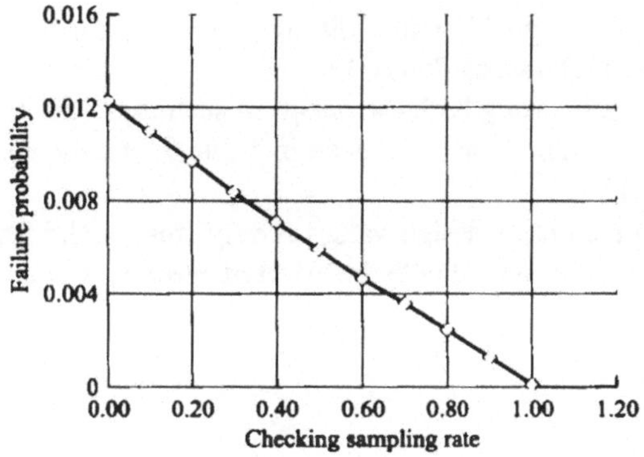

Figure 9. Effect of checking sampling rate in the first checking.

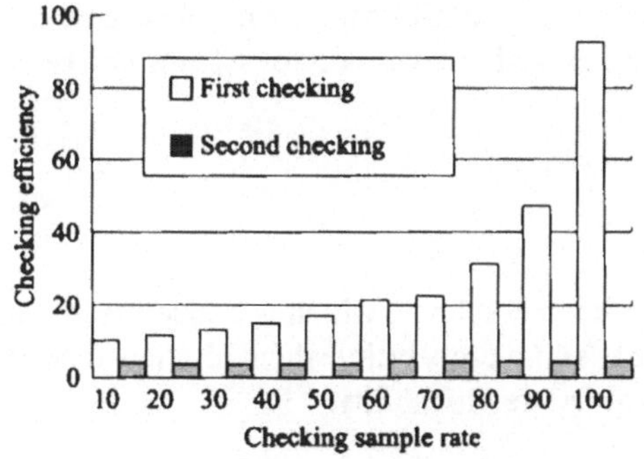

Figure 10. Change of checking efficiency.

Table 7. Failure probability results under different sampling rates.

Sampling rate	Remain conservative error rate	Remain destructive error rate	Remain serious error rate	Total error rate
0.0	0.02404	0.045235	0.004465	0.07374
0.1	0.0219	0.04102	0.00408	0.067
0.2	0.019625	0.0367	0.00365	0.059975
0.3	0.017395	0.032595	0.00316	0.05315
0.4	0.015195	0.02828	0.00275	0.046225
0.5	0.013015	0.02416	0.002315	0.03945
0.6	0.01065	0.019765	0.001845	0.03626
0.7	0.0083	0.015585	0.001365	0.02525
0.8	0.005815	0.01143	0.0009	0.018145
0.9	0.00352	0.00734	0.00039	0.01125
1.0	0.001255	0.00302	0.00001	0.004285

changed. It means that the 100% sampling rate of first checking is very important to a efficacious control of human errors.

4.3 *Results of increasing structural parameter values*

To compensate the destructive effects of human errors to a structural parameter, increasing the parameter values is one of the realistic measures. The failure probability calculated from increasing the designed section height h with 1% under different checking times are listed in Table 8. The corresponding curve is shown as Figure 11.

The failure probability calculated from increasing both the designed section height and the sectional area of reinforcement with checking twice are listed in Table 9, and the corresponding relations are shown in Figure 12.

It is found that the effect of increasing parameter design values is very small under less checking times and this effect is relative small compared with the effect of checking times.

5 OPTIMAL SCHEME OF CONTROL

An optimal control scheme of human errors must be a scheme that is efficient to achieve a required structural reliability at first and to save the control cost. A control scheme may consist of many control measures. The requirements of optimal control scheme can be described as follows.

$$\begin{aligned} &\text{Min} \quad g(m_1, m_2, \ldots m_i, \ldots m_n) \\ &\text{s.t} \quad f(m_1, m_2, \ldots m_i, \ldots m_n) = P_{fcr} \end{aligned} \tag{15}$$

where, $g(m_1, m_2, \ldots m_i, \ldots m_n)$ is the cost of a control scheme which is a function of control measures m_i, $f(m_1, m_2, \ldots m_i, \ldots m_n)$ is the failure probability which also is a function of control measures, and P_{fcr} is the ideal control failure probability.

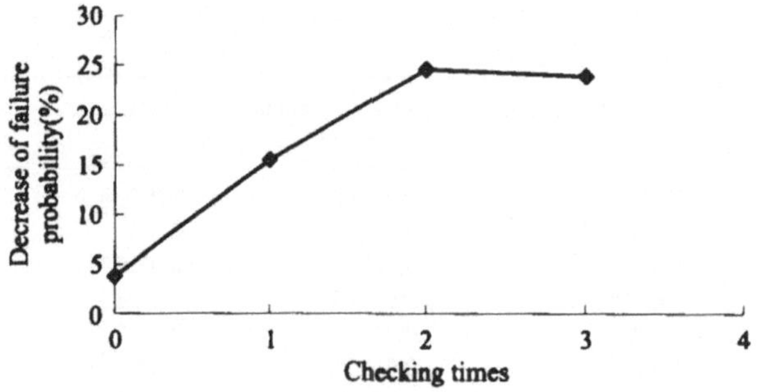

Figure 11. Influence of increasing section height with different checking times.

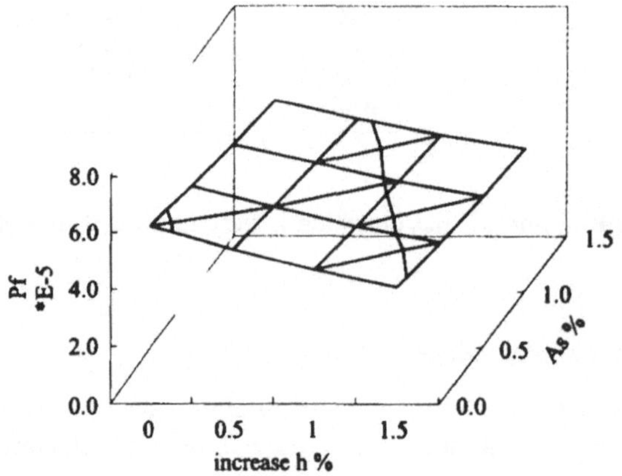

Figure 12. Effect of parameter value (checking twice).

Table 8. Influence of increasing section height under different checking times.

Checking times	Failure probability with original value	Increase value of section height 1%	Decrease of failure probability (%)
0	0.0122659	0.011795	3.840729
1	1.01E–04	8.50E–05	15.53522
2	6.23E–05	4.70E–05	24.5807
3	5.38E–05	4.10E–05	23.85571

Table 9. Results of increase structural parameter value (checking twice).

Increasing area of reinforcement As	Increasing designed section height h			
	0	0.5%	1%	1.5%
0	6.23E–05	5.41E–05	4.70E–05	4.08E–05
0.5%	5.67E–05	4.92E–05	4.27E–05	3.71E–05
1%	5.17E–05	4.49E–05	3.89E–05	3.38E–05
1.5%	4.72E–05	4.09E–05	3.55E–05	3.08E–05

The ideal control failure probability is described as

$$P_{fcr} = \min(P_{fc,} \cdot \gamma P_{fd}) \tag{16}$$

where, P_{fc} is the allowed maximum failure probability which can be found from design code, P_{fd} is the calculated failure probability without considering influence of human errors, γ is a acceptable enlarge coefficient of designed failure probability (assume to be 1.1 in the present paper).

To obtain the theoretical optimal control scheme by Eq. (15), the cost function and the structural failure function must be known at first. This is truly a very complex problem. In other words, it is even impossible to have this two functions for a complex structure. To solve this problem, a simplified method can be used. Some feasible control schemes by some simple analyses can be obtained first, and then analyze and adjust the obtained schemes to choose a relative satisfied scheme, which can be descried as follows:

Step 1: deciding the ideal control failure probability P_{fcr};
Step 2: choosing the possible control measures;
Step 3: analyzing the control efficiency of different control measures;
Step 4: prepare a group of contol schemes and calculating their effects;
Step 5: choosing the schemes that can achieve the ideal control failure probability and analyzing their costs;
Step 6: choosing the least cost scheme as the optimal control scheme.

5.1 *An example of optimal control scheme*

For a rectangular reinforced concrete beam section (Fig. 13) which structural parameters are listed in Table 10, the relative satisfied control scheme explained as follows.

For this section, the failure probability calculated from Table 10 is 3.19×10^{-5}, and the corresponding design code required failure probability is about 6.871×10^{-4} (since for the design code of concrete structures in China, the required safety index β is 3.2 for this type of sections). Thus the ideal control failure probability is determined as

$$P_{fcr} = \min(6.871 \times 10^{-4}, 1.1 \times 3.19 \times 10^{-5}) = 3.51 \times 10^{-5}$$

According to the mentioned analyses, the first round checking is essential for human error control. Based on the first round checking, three feasible control schemes are considered.
Scheme 1: Three rounds checking. The failure probability of this scheme is 5.38×10^{-5}. It excesses the ideal control failure probability, though it satisfies the requirement of design code.
Scheme 2: Two rounds checking with a detailed process control of concrete quality. The failure probability of this scheme is 3.20×10^{-5}. It satisfies the ideal control failure probability.
Scheme 3: Increasing the designed parameter values and two rounds checking. The failure probability control results are listed in Table 11. It is shown that it needs increasing the parameters at least 1% to meet the requirement of ideal control failure probability.

The costs of both Scheme 1 and Scheme 3 are more expensive than that of Scheme 2. It is obvious that the process control of operation is, as in scheme 2, more efficient.

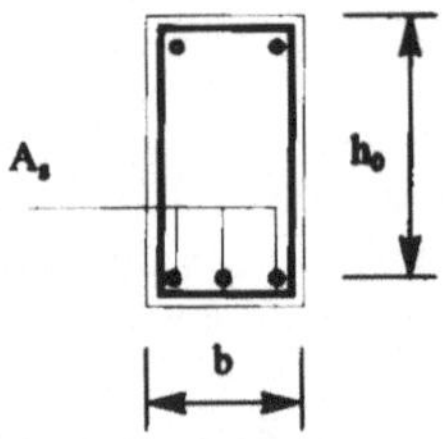

Figure 13. Beam section.

Table 10. Structural parameters without influence of human errors.

Parameters	Distribution type	Expect value	Variance coefficient
Section width b	Normal	250 mm	0.02
Effective section height h_0	Normal	465 mm	0.03
Area of reinforced steel A_S	Normal	1256 mm^2	0.03
Strength of reinforced steel f_y	Normal	374 N/mm^2	0.09
Strength of concrete f_{cm}	Normal	19.6 N/mm^2	0.19
Moment under dead load M_D	Normal	65.3 KN-m	0.07
Moment under live load M_L	Extremum I type	43.1 KN-m	0.29

Table 11. Failure probability results of Scheme 3.

	Increasing h_0			
Increasing A_s	0	0.5%	1%	1.50%
0	6.23E–05	5.41E–05	4.70E–05	4.08E–05
0.5%	5.67E–05	4.92E–05	4.27E–05	3.71E–05
1%	5.17E–05	4.49E–05	3.89E–05	3.38E–05
1.5%	4.72E–05	4.09E–05	3.55E–05	3.08E–05

6 SENSITIVITY ANALYSIS AND DETERMINATION OF KEY CONTROL PARAMETERS

Structural resistance R is a function of several structural parameters and described as

$$R = R(x_1, x_2, ..., x_n) \tag{17}$$

Define a parameter as

$$\lambda_i = \frac{\partial R}{\partial x} \cdot \frac{x_i}{R} \tag{18}$$

This parameter reflects the sensitivity of structural resistance to parameter x_i dispersive from its designed value, which is called sensitivity factor. The bigger the sensitivity factor is, the more possible of the human error to influence the structural parameter and to cause structural failure.

Figure 14 shows the sensitivity factor curves of rectangular section of a reinforced concrete beam, which shows that the sensitivity of structural parameters with the relative compression depth of the beam section ξ.

The designed structural resistance usually should have a bigger safety margin compared with the requirement of design code. In this case, very small parameter dispersion from its designed value due to the effect of human errors will not cause structural failure. In fact, the bigger margin has decided the maximum tolerance of the structure towards human errors. A concept to describe the tolerance is presented here. If human error occurred, it may causes a structural parameter dispersive from its designed value and reduce the structural resistance. When the effect magnitude of human errors reaches a certain value, the practical structural resistance just equals the code required resistance, this human error magnitude is so called at critical point, which is illustrated as Figure 15.

The critical point can be obtained from a series of structural analyses through changing the human error effect magnitude.

The occurrence rate of human errors is a function of the effect magnitude. Besides the critical point, the human error occurrence possibility is also very important for deciding the key parameters in human error control. It is reasonable to combine the critical point and the occurrence possibility of human error that influence the decision making of the control measures. This idea is illustrated in Figure 16. In this illustration, the decision can be made according to the possibility zones of human error occurrence. The classification of the human occurrence possible zones can be adjusted, which depends on the practical need. Table 12 is the proposed possibility classification of human error occurrences.

According to the possibility zones of human error occurrence, three control measures are proposed as follows.

Case 1: If the critical point is located in the high possible zone, the parameter is a key parameter, and two rounds detailed checking must be performed. If the key parameter is associated with concrete quality, a detailed process control must be adopted.

Case 2: If the critical point is located in the possible zone, the parameter is an important parameter, two rounds checking must be performed.

Case 3: If the critical point is located in the low possible zone, the parameter is a non-important parameter, and only a general checking is needed.

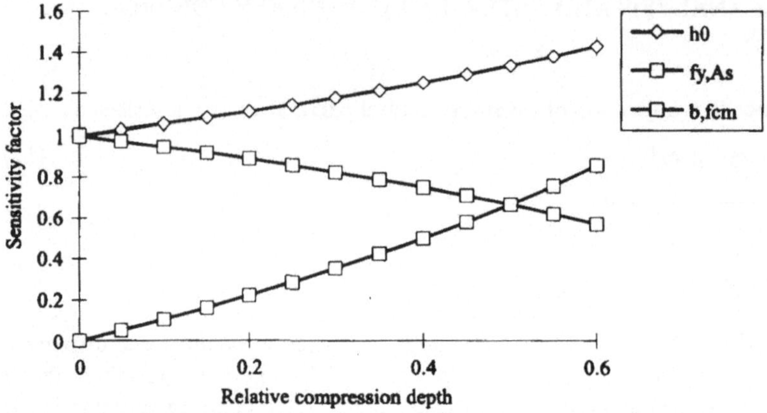

Figure 14. Sensitivity factor of a beam section.

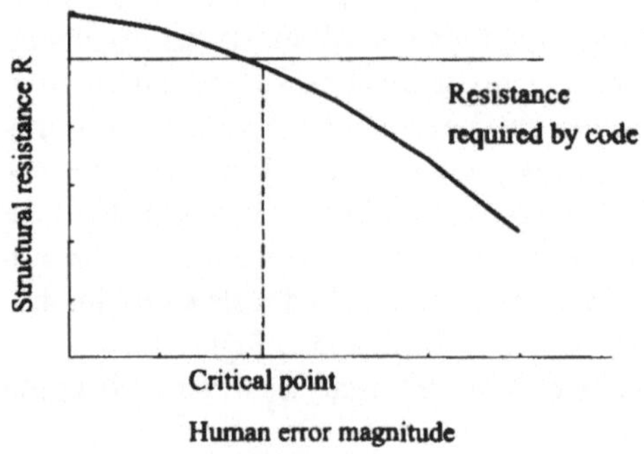

Figure 15. Illustration of critical point.

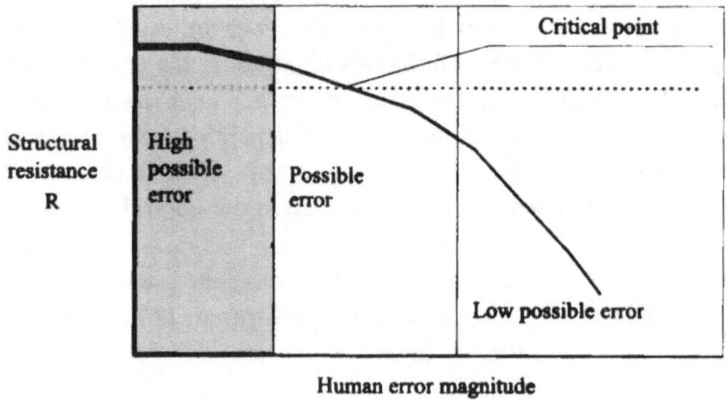

Figure 16. Illustration of choosing key control parameter.

7 SAFETY ANALYSIS AND CONTROL OF STRUCTURE DURING CONSTRUCTION

The construction stage is very important for structural safety. In this stage, the structure is in fact a time-dependent structure and is more complex than that of the normal use stage. The loads during construction are sometimes larger than those considered in design. Since all the structural parameters are determined through construction operations, the safety analysis and control of a structure during construction is necessary. The safety control considered during construction must include to control both safety of construction and safety of performance. In construction stage there are many construction steps. The analysis and control process of structural safety during a typical construction step can be illustrated as Figure 17.

8 A REINFORCED CONCRETE FRAME ILLUSTRATION

A reinforced concrete frame with two-stories and two-spans is analyzed to illustrate the mentioned method to choose an efficacious scheme for human error control. The frame is shown in Figure 18.

Table 12. Classification of human error occurrences.

Classification	Occurrence possibility of human errors	Human error magnitude	
		Lower bound	Upper bound
High possible error	> 0.01	0.0	a
Possible error	0.001–0.01	a	b
Low possible error	< 0.001	b	1.0

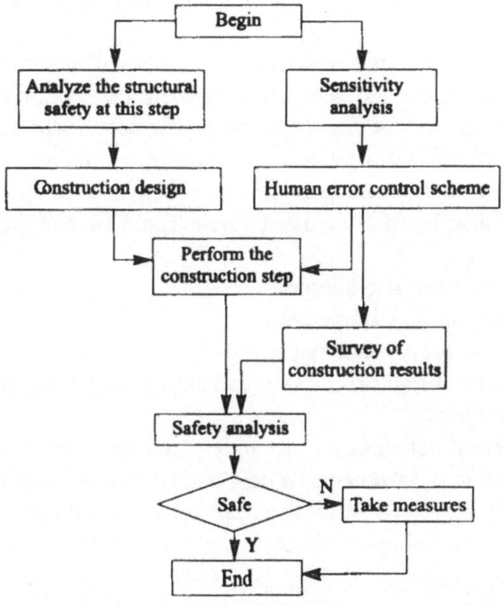

Figure 17. Safety analysis and control during construction.

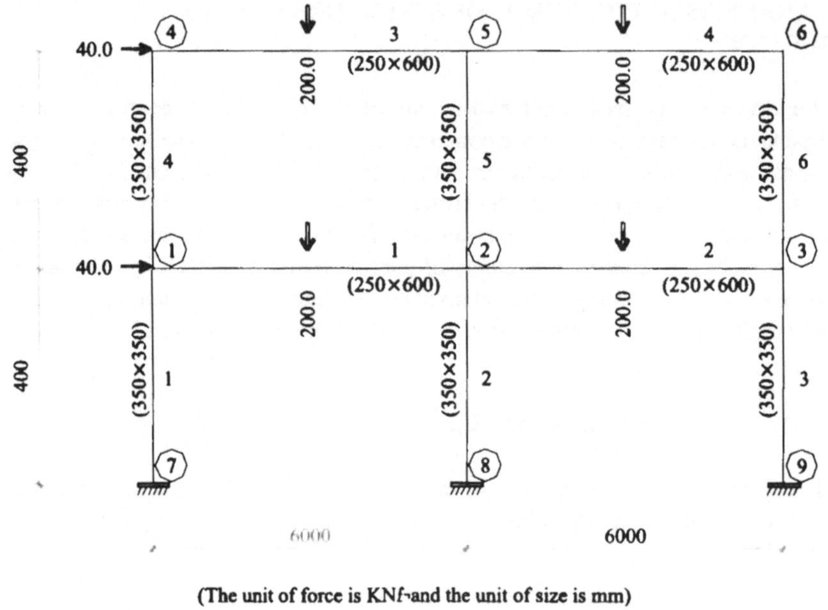

Figure 18. A reinforced concrete frame.

To illustrate the method, only the final stage during construction is considered. An evaluation index of system reliability that take the member functions into account is used to evaluate the control efficiency (Xu Maobo etc.1997).

The classification of the human occurrence possible zone according to the Table 12 is obtained from the simulation results and is adjusted according to advice from some experts. The classification is as Table 13.

Through the system structural analysis, the required resisitance of structural member sections can be obtained. According to the results of sensitivity analysis on sectional resistances, the critical points of human errors for the main structural parameters can be obtained, which are listed in Table 14. For each main section, a control measure can be chosen by the obtained critical points (Table 14) and possible zone human errors of Table 13. The control measures chosen are also listed in the last column Table 14, and the proposed control scheme has been done.

For comparison, three other control schemes are also adopted.

Scheme 1: performing one round checking for all parameters.

Scheme 2: performing two rounds checking for all parameters.

Scheme 3: performing two rounds checking for all parameters, and performing detailed process control for concrete quality at all sections.

To compare the control cost among different schemes, a cost unit is defined here for simplicity. One round checking of one section is assumed as a unit cost of control, and a detailed process control of concrete quality for one section is assumed as two cost unit. The total cost of a control scheme can be obtained by accumulate the cost for all control sections.

Table 13. Human error magnitudes in different zones (%).

Parameters	High possible errors	Possible errors	Low possible errors
Strength of concrete	< 15	15–50	≥ 50
Area of reinforcement	5	5–40	40
Effective section height	5	5–40	40
Section width	2	2–15	15

Table 14. Critical points of human errors for different structural parameters (%).

Position	Parameters				Control measure
	Strength of concrete	Strength of steel	Effective section height	Stirrup steel	
1 – *left*	–	–	–	–	C
1 – *midspan*	19.5	3.8	2.9	–	A
1 – *right*	14.8	5.4	7.6	39.0	A
2 – *left*	52.9	46.3	34.5	–	C
2 – *midspan*	46.4	13.5	10.5	–	B
2 – *right*	39.5	7.9	6.4	–	B
3 – *left*	–	–	–	–	C
3 – *midspan*	15.3	2.8	2.2	–	A
3 – *right*	27.1	12.5	12.0	44.8	B
4 – *left*	40.2	29.3	23.2	–	B
4 – *midspan*	30.0	6.7	5.1	–	B
4 – *right*	–	50.2	42.8	–	C
① – *upper*	–	52.9	25.2	–	C
① – *bottom*	–	–	54.2	–	C
② – *upper*	30.2	49.1	12.4	–	B
② – *bottom*	34.8	58.4	14.8	–	B
③ – *upper*	–	39.3	20.9	–	B
③ – *bottom*	–	31.4	17.2	–	B
④ – *upper*	–	43.9	26.0	–	C
④ – *bottom*	–	43.7	26.7	–	C
⑤ – *upper*	–	–	32.9	–	C
⑤ – *bottom*	–	–	31.1	–	C
⑥ – *upper*	–	50.3	32.6	–	C
⑥ – *bottom*	–	5.3	3.5	–	A

Annotation:
The critical points over 0.6 are not listed in this Table.
Measure A: two rounds detailed checking and detailed process control of concrete operation.
Measure B: two rounds checking.
Measure C: one round checking.

The calculated results of the structural safety index under different control schemes are listed in Table 15. The total control costs of these control schemes are also listed in.

According to the investigation, the control scheme 1 roughly corresponds with the practical control level in China now. It is found that the control result of the scheme is not acceptable. Compared with the other two control schemes, the proposed scheme that obtained from the above analyses has a best control result and it costs less.

Table 15. Comparison of control schemes.

Control schemes	Functional damage of the structure	Total control cost (cost unit)
Designed level (error-free)	8.77×10^{-6}	
proposed scheme	1.28×10^{-5}	41
Scheme 1	2.76×10^{-5}	24
Scheme 2	1.61×10^{-5}	48
Scheme 3	8.79×10^{-6}	96

9 CONCLUSIONS

The present simulation model for that the human error rate is reasonable, in which effect magnitude of human error and human error checking process are all taken into account. The simulating results have good consistency with the investigating results.

The proposed calculating method on structural failure takes human error into account and it can also be used to analyze the very complex case in which combined human errors occurred.

It is more reasonable to choose the key control points by considering both the structural sensitivity to human errors and the occurrence possibility of human errors, by which an applicable method is presented.

A satisfied control scheme can be obtained using control efficiency analysis. It is illustrated from the analysis of an example that the results seem reasonable.

REFERENCES

Xila Liu 1997. *The status and prospects of structural engineering*. The people's transportation publisher (in Chinese).

Xu Maobo 1998. Analysis and control of structural safety considering the human errors during construction, Dissertation submitted to Tsinghua University for the degree of doctor of engineering (in Chinese).

Nessim, M.A. & Jordaan, I.J. 1983. Decision-making for error control in structural engineering. *Fourth International Conference on Applications of Statistics and Probability in Soil and Structural Engineering*, Florence, 1983.

Nowak, S. & Lind N.C. 1985. Modeling human errors, *IABSE Symp. On Safety and Quality Assurance in Civil Engineering, IASBE Report*. 1985, Vol. 50.

Stewart, M.A. & Melchers R.E. 1988. 'Simulation of human error in a design loading task', *Structural Safety* 5(4).

Stewart, M.A. Probabilistic risk assessment of quality control and quality assurance in structural design, IEEE Transactions on Systems, *Man and Cybernetics* 21(5).

Swain, D. & Guttman, H.E. Handbook of human reliability analysis with emphasis on nuclear power plant applications, NUREG/CR-1278, US Nuclear Regulatory Commission, Washington, D.C.

Apostolakis G. 1982. Data analysis in risk assessments. *Nuclear Engineering and Design*, Vol. 71.

Stewart M.G. 1993. 'Modeling human performance in reinforced concrete beam construction', *Journal of construction Engineering and Management* 119(1).

Park K.S. 1987. *Human reliability*, Elsevier.

Xu Maobo, Zhao Jinkai et al. 1996. 'Human errors during construction of reinforced concrete structural and their influence to concrete strength', Engineering mechanics (in chinese), Supplement.

Xu Maobo, Liu Xila & Lee Huiming 1997. 'A new method of structural system reliability evaluation', *Journal of structural engineering* (in Chinese) (4).

CHAPTER 5

Health risk management in construction

MICHAEL N. BAXLEY

ABSTRACT: Health risk management in construction poses unique challenges due to the nature of the industry. Occupational injuries and associated costs are disproportionately represented in this small segment of total working population. Although occupationally related illnesses are well documented, they are difficult to recognize in the general practice of medicine and temporal management of the construction workers' health, generally permanent in scope and largely preventable. Therefore a strong focus on prevention, safety and early recognition is the key to optimal health risk management.

Health risk in the working population is extremely broad and probably largely unrecognized. Surveys of hospitalized patent indicated that 75% report some form of hazardous exposure. Approximately 17% of all workers suspect a link of their illness to the job or about 860,000 illnesses per year. Moreover it is estimated that about 60,300 deaths occur per year in the working population secondary to some aspect of occupational exposure.

In the construction industry, which represents 13% of the Gross National Product with only 5-6% of the labor force, there appears to be a disproportionate injury rate in regards the overall working population. The construction working force suffers 15% of all occupational fatalities, with 17% of all workers' compensation costs and 9% of lost work days. This accounts for around 10-40 billion dollars in the U.S. If one sets the cost at an average of 20 billion dollars, the cost per construction worker is around $3500 per year. In fact the workers' compensation premiums for three trades – carpenters, masons, and structural iron workers – accounts for approximately $28 for every $100 payroll. This does not take into account the indirect costs associated with workers' compensation injuries and illness such as liability premiums, reduced work efficiency, overtime and clean up costs.

Despite its high rate of injury/illness, most of the costs are associated with predictable ailments. On the injury side, these are the trauma from falls seen usually in the roofer, steelworkers or laborers; the occupational white finger seen in the jack hammer operators, lower back pain in the bricklayers, carpal tunnel syndrome in the iron workers or electricians. Probably less well defined and less obvious to the uninitiated medical practitioner, but also very predictable are the occupational illnesses related to construction: pneumoconiosis in coal workers, tunnel builders and welders, skin allergies in the mason, kidney ailments in the painter and roofer (secondary to potential for exposure to solvents), asbestosis in building demolition, lead poisoning in bridge rehabilitation, and heat stress in the hazardous waste worker. The most important things to remember here is the preventable nature of these processes and the permanence of occupational illnesses in general.

The construction industry unfortunately lends itself to a grim picture in safety and health because of the way it operates. Construction rarely provides steady employment and it is hard to monitor the work history of the employee, his injuries, or his use of personal protective equipment or exposure to hazards. Construction sites allow and require movement of the worker from site to site. The work assignments change, as well as the topography and the number of trades at the work site. In fact the workers are always working themselves out of a job, and may have several employers per year. Most contractors are self-employed. And 80 to 90% have fewer than 10 employees. However they still only manage about 1500-2000 hours working time per year, and much of this may be in lumped segments of time. Many hours of overtime interspersed with long stretches of no work available or having to travel long distance for employment opportunities may actually pose a long term risk of stress. Many of the workers are situated in union arrangements which further adds to the inequity of the industry – 7 million people in construction yet they are equivalent to only 4-5 million full time employees in the other industries.

Adding to the difficulty posed by the nature of the industry are direct and indirect consequences of the system 'designed' to take care of them. Many employees, employers, and providers and mangers of health care have difficulties in dealing with the workers' compensation system. More recently the managed care environment has placed further restraints on the systems reducing the time available to take a complete occupational history, follow up on injury or illness or interact with the employer through he intermediary insurance carrier or third party payor.

Many employees may be reluctant to connect a health problem with their work for fear of losing their job, or failure to recognize a temporal association. Add this to the practitioners potential for under recognition of occupational illness by failing to make connections with the work environment may interfere with effective health management. In summary, these processes created in an effort to control the cost of the injury, may have effected the attention given to recognition, prevention and intervention. This has created a dichotomy of function between workers' compensation and its misapplied pseudonym 'Occupational Medicine.'

In European countries where the management of workers' injuries has matured inside the delivery system of occupational medicine many interventional prevention measures have been applied at several levels. In the Scandinavian counties as well as Germany, chromate induced dermatitis has been addressed by adding complexing agents to the cement in final stages of preparation; Cement block with handholds have reduced lower back injuries in Germany; mandated hearing protection in Sweden has reduced the incidence of occupationally related hearing loss; rod tying process reworking has allowed construction workers to remain upright while performing this task; use of wet sand blasting in Europe has reduced the risk of silicosis; modification of the handles in screwdrivers reduced work related wrist injuries; slip resistant boots have reduced falls in Canada; and changing the size of wall board in the Scandinavian countries to allow ease of handling have reduced upper extremities and lower back strain. A proactive preventive approach mandated by the governmental regulators of the industry which is applied by the employers and health risk is minimized and therefore maximally managed in a cost effective manner.

What we see here is a developing pattern or an argument of structure. Donald Millar, National Institute of Safety and Health (NIOSH) Director, in 1996 said '*'The biggest oc-*

cupational health and safety problem is the blindness of our society to the benefits of prevention,' This is best understood by applying the principles of prevention in the management of health risk in general. Prevention can occur at three levels of intervention: Primary, Secondary, and Tertiary. This can be illustrated graphically in Figure 1 by examining the relationship between impairment and disability. In this approach the basic man-environmental relationships are exemplified by impairment, the precursor of disease. Not withstanding environmental stressors outside work, cost (energy expended, currency) is determined by increasing loss of function or increasing impairment (person, system) and is directly proportional to work stressors regardless of etiology.

Operating at the primary level is the least expensive with the greatest return on investment. This is the level where true occupationally oriented prevention programs are applied. This is also the level where safety and health make their biggest impact, before an injury or illness has occurred. The system or worker, maintains homeostasis for the least cost in energy or dollars. In the beginnings homeostasis is maintained with normal adjustment.

At the secondary level of prevention, an injury or illness has already occurred, and efforts are now directed at prevention of permanent impairment or disability. Normal function is maintained without significant cost; yet cost increases with movement along the compensation line. This is the level at which most Workers' compensation injuries are managed. Once a worker falls into this category of impairment, occupationally related illnesses and injuries must be identified and attempts at treatment rendered. The goal here is to treat early and effectively and return the system or worker back to full function.

Finally, there is the tertiary level of prevention. Prevention of major disability in the face of permanent loss of function is attempted. The worker may have reached the limits of his body's compensatory processes, and intervention ins necessary to effect change and maintenance of function. This is essentially disability management. Preservation of function is paramount, however the returns on investment are much less pronounced. Expenditures are generally higher and results less encouraging. Nevertheless the further back on the curve one is able to achieve these goals, the more predictable and long lasting the results. Early recognition and management is important, and is not limited just to medical management.

Health Risk Management therefore is best addressed as modifiable factors in a construction worker's environment and system. These include Labor/Management Site Safety and Health Plans; Hazard Identification and Monitoring, Training and orienting of workers and supervisors (including Hazardous Communication); Substitution of Equipment, Materials and tools; Use of Personal Protective Equipment; Medical Testing as needed; Record Keeping and reporting; Effective Worker's Compensation Systems; Effective Occupational Health Delivery Systems and Capitalizing on Public Health Opportunities – Improved Preventive Medicine, Improved Disability Determination, Improved Rehabilitation, Research including Data Generation.

Applied industry wide, as evidenced in the earlier examples, an effective safety and health plan would result in reduced accidents and injuries, increased efficiency, improved morale, less downtime, and higher quality products. At the local level the effects may be more profound: Reduced accidents and injuries increased efficiency, improved morale, less downtime, higher quality products. Moreover, locally the programs will show economic impact as illustrated by using the Experience Modification Rate (EMR). This derived figure is directly proportional to the lost-time injury rate. In equation format this is (hourly wage) × (manual rate) × (EMR) = hourly workers compensation cost per worker. Businesses with higher compensation rates have higher costs of doing business.

Hazard identification and monitoring requires application at the management level with knowledgeable input from the Occupational Health program responsible for worker health management. Training of workers and supervisors is the key to effective dissemination of risk. This includes additional instruction in the proper use of personal protective equipment, and judicious ergonomically and economically directed substitution of tools and processes.

Labor management cooperation is equally important in health risk management in construction. The worker needs to understand the importance of these programs and when possible participate and monitor the performance of the programs. These teams could be assembled from supervisors and site foremen with direct input from the workers. Guidelines for the workers could be developed as well as appropriate incentives for the participants of safety programs – such as Safety recognition awards. The ultimate goal is the elimination of critical risk behaviors in the worker. These often arise when the worker attempts to shed a temporary inconvenience in use of personal protective equipment, proper body mechanics or materials handling. All could contribute to injury or illness – all are preventable.

Safety and Health Hazards for the constriction worker are well documented. Table 1 outlines the causes of the most common category of injury – Musculoskeletal – as well as the areas most effected. Table 2 outlines the most common toxic hazards. Less recognized but none the less contributory to overall potential for temporary or even permanent dysfunction the Chronic Health Hazards and the organs most effected listed in Table 3. All of these conditions and stressors are preventable, either at macro or micro level, but are not fully achievable by any single medical or non-medical system operating alone.

Until a national or industry wide approach to such programs is achievable, local control of the aforementioned modifiable factors is needed. This can be greatly aided by effective partnering with a reliable occupational health practice that provides a full scope of occupational health services in addition to effective workers' compensation management of illness and injury. Occupational medicine by definition is the study, management, monitoring, and prevention of work related illness and injuries. It is the intersection between curative medicine and preventive medicine, and the science of understanding the health effects of man's interaction with his environment, whether work or non-work related.

Full service occupational health programs not only provide optimal workers' injury and illness management, they also provide a host of necessary services that are related to overall prevention and monitoring in workers' health. They are capable of treating one patient for one visit for one injury; however they are also capable of health maintenance and monitoring, risk identification processes such as physical examinations, audiometric,

Table 1. Physical hazards.

Musculoskeletal Disorders Causes	Area(s)
Lifting	Lower back, shoulders
Awkward postures	Knee, hip, shoulders
Repetitive motion	Wrists
Hand-tool vibration	Fingers, wrists

Table 2. Toxic hazards.

Dusts	
Asbestos	Demolition, maintenance, insulation
Cement	Foundations, sidewalks, floors
Fiberglass	Other insulation pipe insulation, air conditioning
Silica	Sandblasting, tunneling
Wood dust	Remodeling, demolition, sawing
Metals (dusts and fumes)	
Cadmium	Welding, cutting pipe
Hexavalent Chromium	Welding, cutting pipe
Copper	Welding, cutting pipe
Lead	Demolition, lead paint surfaces
Magnesium	Welding, cutting pipe
Zinc	Welding, cutting pipe
Solvents	
Benzene	Hazardous waste cleanup, Petrochemical plant sites
Methylene Chloride	Paint strippers
Toulene	Varnishes, paints, adhesives, cleaners
Trichlorethylene	Varnishes, paints, adhesives, cleaners
Other Chemicals	
Epoxy Resins	Impermeable paints, wood floor primers
Polyurethanes (isocyanates)	Seam sealers, insulation, electric wire coating
Coal Tar Pitch	Roofing, road work

Table 3. Chronic health hazards.

Hazard	Organ/System
Noise	Hearing
Silica, asbestos, manmade fibers	Lungs
Lead and other metals	Kidneys, nervous and reproductive systems
Solvents	Kidneys, liver, nervous system
Hazardous wastes	All systems
Heat and extreme cold	Circulatory system

and spirometric examinations, and fitness for duty and respirator screening. All programs should provide medical testing as needed and adhere to meticulous Record Keeping and reporting practices. They should also should aim for improved Preventive Medicine, Improved Disability Determination, and Improved Rehabilitation – all directed returning ht worker to function and to work.

More advanced programs may have additional services such as blood borne pathogen training, drug screening programs, medical review officer services, and occupational health consultation on a variety of industrial issues. They may also have the capability of research and data generation. Regardless though of scope of function their should be a through understanding of Worker's Compensation Systems and efforts made at capitalizing on Public Health Opportunities.

A truly effective occupational health and workers' injury management program will operate under consistent quality protocols and have accessibility during work hours. Although the program doesn't have to be a wholly occupationally oriented clinical practice with a provider with specialization in occupational medicine, its purveyors of service must show an interest in the worker vis-à-vis the workers' environments. This includes at a minimum a willingness to work with the workers management and safety supervisors, and fully understand the workplace., Research including Data Generation.

Operationally the construction worker is best evaluated and managed by knowledge of the work.. This best accomplished by a good occupational survey. Table 4 lists six screening questions that address the worker's health and work exposure status. A more compete evaluation includes a detailed Occupational History (Table 5) focusing on jobs, exposures, symptoms at work or home, and also a knowledge of any non-work related activities or exposures that may contribute to the issue at hand. Finally, a number of resources (Table 6) exist to aid not only the Occupational Health Provider, but also the worker's management in identification of potential hazards.

Table 4. Key occupational medicine screening questions.

1. What type of work do you do?
2. Do you think your health problems might be related to your work?
3. Are your symptoms different at work and at home?
4. Are you currently exposed to chemicals, dusts, metals, radiation, noise or repetitive work in past?
5. Are any of your co-workers experiencing similar symptoms?

If the answers to one or more of these questions suggest that a patient's symptoms are job related or that patient has been exposed to hazardous material, a comprehensive occupational history should history should history should be obtained.

Table 5. Elements of occupational history.

List of Jobs	Lifetime history, with dates of employment and job duties Military history
Exposures	
Type	Chemicals Metals Dusts Biologic Physical
Assessment of Dose	Duration of exposure Exposure concentration Route of exposure Presence and efficacy of controls Quantitative exposure data from inspections and monitoring
Timing of symptoms in relation to work	Symptoms exacerbated or occur at work and improve away from work Symptoms coincide with the introduction of new exposure at work or other change in working conditions Presence of similar symptoms among co-workers with the same type of job and/or exposures
Evaluation of non-work exposures	Home environment Hobbies or recreational activities

Table 6. Occupational safety and health resources.

Association of Occupational and Environmental Clinics (AOEC) More than 50 mostly academically based clinics belong to the AOEC. Most clinics employ multidisciplinary staffs for clinical and work place evaluations and consultations. For a listing of clinics, contact the AOEC: Association of Occupational and Environmental Clinics 1010 Vermont Avenue Suite 513 Washington, D.C. 20005 Telephone: 202-347-4976 Internet address: http://152.3.65.120/OEM/AOEC.htm
American College Occupational and Environmental Medicine (ACOEM) The ACOEM can also be contacted for a list of occupational medicine clinical resources: American College of Occupational and Environmental Medicine 55 W. Seegers Rd. Arlington Heights, {L 60005 Telephone: 847 6850 Internet address: http://www.acoem.org
National Institute for Occupational Safety and Health (NIOSH) NIOSH is the national agency mandated to research occupational hazards, carry out workplace health evaluations and fund the training of occupational safety and health professionals. NIOSH is an important source of information on specific hazards and a wide range of issues related to occupational safety and health: National Institute for Occupational Safety and Health 4676 Columbia Pkwy. Cincinnati, OH 45226 Telephone: 800 356 4674 Internet address http://www.cdc.gov
Occupational Safety and Health Administration (OSHA) OSHA sets and regulates workplace exposure standards. OSHA maintains offices in many locations across the country. Physicians may contact OSHA for information or file complaints with the agency. The address of the local OSHA office can be found in the telephone book in the federal government listings, under the Department of Labor: Occupational Safety and Health Administration 200 Constitution Ave., N.W. Washington, DC 20210 Internet address: http://osha.gov
Poison control center Poison control centers are located across the country. These centers can be consulted for information on treatment of overexposures and for information on various hazards as obtained in large databases: Internet address. http://cdc.gov
State and local health departments State and local health departments vary in the level of occupational safety and health resources they employ. Some departments are important sources of information and referral. A few conduct workplace investigations.
Computerized databases A variety of computerized databases are available. Medical libraries should be able to access the National Library of Medicine Toxicology Data Network (TOXNET). This network include a number of files that allow a user to research the effects of archives exposure.

Table 6. Continued.

Compact disc read only memory (CD-ROM) databases; CD ROM databases are available from a number of source. They can be purchased as a subscription, with regular updates. Micromedex, Silver Platter and the Canadian Center for Occupational Health and Safety are some of the most commonly used CD-ROM systems.	

Source: Lax, et al., AMERICAN FAMILY PHYSICIAN, SEPTEMBER 15, 1998, VOLUME 58, NUMBER 4, page 943

In summary, health risk management in construction is dependent on a number of variables working in tandem to achieve a common goal. Prevention of the injury is paramount and involves labor and management attention to safety and health protocols. But prompt recognition of illness and effective treatment by health care providers in conjunction with strong preventive medicine approaches also plays an important role. Together they result in an integrated systems approach to overall health risk management in the construction worker.

REFERENCE

Leigh, J.P., Markowitz, S.B., Fahs M., Landrigan, P.J. 1997. Occupational injury and illness in the United States. Estimates of Costs, morbidity, and mortality. Archives of Internal Medicine; 157: 1557-1568.

Rigen, K., Seegal, J., Englund, A. 1995. Safety and Health in the Construction Industry, Annual Review of Public Health 16: 165-88.

Rigen, K., Stafford, E. Intervention Research in Occupational Safety and Health: Examples form Construction, *American Journal of Industrial Medicine* 29: 314-320.

Lax, M., Grant, W., Manetti, F., Klein, R. Recognizing Occupational Disease – Taking an Effective History, American Family Physician 58(4): 935-944.

Hatch, T.F. 1962 Changing Objectives in Occupational Health, *American Industrial Hygiene Association Journal* 23:1.

CHAPTER 6

Principles of exposure assessment – application and experience with a task-based approach

PAM SUSI

ABSTRACT: Construction workers experience disproportionate mortality associated with occupational diseases such as asbestosis. The degree to which workers are at risk of disease and premature death is dependent on their exposure to hazardous agents. Measuring occupational exposures in construction is particularly challenging since variables which effect exposure, such as the worksite, the workforce, and the materials in use, are constantly changing. The Center to Protect Workers' Rights (CPWR), the research arm of the Building and Construction Trades Dept., AFL-CIO, has developed and utilized a 'Task-Based Exposure Assessment Model (T-BEAM)' for determining exposure to welding and thermal cutting fumes among construction trades. T-BEAM employs two defining characteristics: 1) involvement of specially trained journeyman workers in the exposure assessment process; and 2) an emphasis on the identification, implementation and evaluation of engineering and work practice controls.

Between 1995 and 1996, CPWR collected 195 full shift personal exposure samples among 63 workers employed in welding trades on 10 construction sites located throughout the US. Samples were analyzed for total particulate. In addition, a smaller number were analyzed for constituent metals such as manganese, nickel and chromium. Probabilities of exceeding occupational exposure limits (OELs) were high for all trades (boilermakers, 100%; ironworkers, 71%; and pipefitters, 31%). The probability of exceeding OELs for manganese were lower but still high for boilermakers (72%). However, very few samples were analyzed for manganese and other constituent metals. Variables which effected exposure were trade, task, whether work was performed indoors or outdoors and use of ventilation. Use of mechanical or local exhaust ventilation reduced exposures significantly. The sample mean for 1996 total particulate exposure measurements associated with use of mechanical and local exhaust ventilation was 1.79 mg/m^3 (22 samples) versus 5.46 mg/m^3 (57 samples) with natural or no ventilation. These data suggest that construction workers employed in the welding trades are likely to be at risk of adverse health effects associated with welding and thermal cutting fumes. Because employers are required by the Occupational Safety and Health Act to provide a 'safe and healthful' workplace, contractors who fail to incorporate engineering controls such as use of ventilation for welding, appear also to e at risk of being cited for violation of current standards which designate Permissible Exposure Limits (PELs) for particulate and specific metal constituents of welding fumes.

Because construction is often performed in industrial environments which present process hazards and multiemployer job sites create an array of trade specific health hazards, a 'systems' approach to controlling hazards is needed. Owners/construction clients can play a key role in ensuring that contractors comply with regulatory and contractual requirements aimed at protecting the construction workforce. In addition, joint safety and health committees with employer and employee participation from each contractor and trade on site will facilitate the necessary communication for identification and control of hazards.

1 INTRODUCTION

Approximately one worker dies of asbestos-related disease eacn hour in the US and the number of deaths is expected to continue to climb until after the year 2002. Construction workers who represent 6-7% of the workforce represent 34% of deaths from asbestos-related cancer (CPWR, 1998). These startling figures illustrate the importance of charactering exposures that cause occupational disease. Exposure assessment is fundamental to the development and evaluation of effective strategies for preventing occupational illness and injury. Health hazards in construction are numerous and encompass physical hazards such as noise, ergonomic hazards and chemical hazards. Some of the more common chemical hazards encountered in construction are natural and man-made mineral fibers, asphalt and coal tar pitch fumes, welding and thermal cutting fumes, and silica dust. A field investigation of these hazards on a new construction project underway between 1991 and 1992 documented several instances in which over exposure to these agents occurred (Schneider & Susi 1993). In addition, the use of engineering and work practice controls to reduce exposures to chemical agents are underutilized in construction. This is due in part to a poor understanding of health risks in construction and exposure levels which give rise to that risk.

Exposure assessment is a primary function of the occupational/industrial hygienist. In order to be effective in this capacity, the practicing occupational hygienist must have a general knowledge of occupational health hazards, principles of hazard control and an understanding of the work performed by the labor force in the industry in which they are employed. Assessment and control of health hazards in construction is uniquely challenging in that the work force is highly mobile and the worksite is perpetually changing. In response to this, researchers have developed a. 'task-based' approach to exposure assessment which seeks to describe occupational exposures among construction workers with emphasis on defining and measuring variables that impact on exposure. This approach, developed and in use by the Center to Protect Workers' Rights-the research arm of the Building and Construction Trades Dept., AFL-CIO, relies heavily on the knowledge and involvement of the journeyman construction workers in the exposure assessment process.

This chapter will present: 1. The basic principles of exposure assessment and how they apply to construction; 2. The occupational safety and health laws which require employers to develop exposure assessment programs; 3. The use and results of a Task-Based Exposure Assessment Model (T-BEAM) for characterizing welding and thermal cutting fume exposures among construction workers; and 4. Unique problems associated with construction work.

1.1 *Principles of exposure assessment in construction*

Mortality and morbidity in construction. Over 1000 workers die on US construction sites each year. The hazardous nature of construction from acute trauma such as falls, electrocutions, trenching cave-ins and other fatal injuries has long overshadowed the more prevalent and insidious impact of occupational disease and impairment resulting from work at the construction trades. For example, NIOSH estimates that over 2.4 million construction workers are exposed to hazardous noise (NIOSH 1998). Noise induced hearing loss manifests itself gradually over a period of years. However, the number of construction workers that experience measurable hearing loss at a relatively young age are alarmingly high. Based on a database composed of over a million hearing tests conducted over a 16 year period, only 20% of carpenters had normal hearing in both ears by age 41 (CPWR 1998).

Noise induced hearing loss, like other occupational illnesses, catalyzes other losses that are harder to measure, but extremely important to the life and well-being of workers and their families. For the worker whose hearing is gradually diminished while still in the prime of their life, decreased hearing may impact on safety and communication with co-workers. In addition, quality of life may be reduced as it becomes increasingly difficult to communicate with loved ones or enjoy activities requiring clear comprehension of language. Other more deadly diseases associated with construction hazards, such as lung cancer, silicosis, asbestosis, and other respiratory diseases have similar effects in that reduced lung capacity may shorten the career of a construction worker and confine them to a less active lifestyle than they would normally enjoy.

In recent decades, several investigators have documented increased morbidity and mortality among construction workers from diseases associated with health hazards common to construction. However, the exposure levels which give rise to disease are less well understood. Such information is vital for disease prevention strategies aimed at controlling occupational health hazards.

Elevated disease rates within trades that could be associated with task-specific exposures signal a need for research aimed at characterizing and reducing exposure risk. For example, boilermakers, an important welding trade, experience elevated mortality rates from lung cancer (Beaumont & Weiss 1980). The International Agency for Research on Cancer has concluded that welding fumes are 'possibly carcinogenic to humans' (IARC 1990). Laborers experience elevated rates of pulmonary tuberculosis, a disease associated with silicosis. They are frequently exposed to silica dust during demolition, concrete and clean-up work. Electricians and boilermakers, trades often employed in power houses and near electromagnetic field sources suffer excess aplastic anemia and acute myeloid leukemia, respectively.

Assessing exposure risk. The degree to which workers are at risk of disease and premature death is dependent on their exposure to hazardous agents. Exposure in turn is dependent on a number of factors including:

- The route through which a hazardous substance enters the body (e.g. inhalation, ingestion or skin absorption);
- The concentration of the agent in the air for inhalation hazards and on surfaces for those hazardous agents for which ingestion is a viable route of exposure; and
- The duration of time in which a worker comes into contact with a hazard.

Occupational health hazards fall into two major categones: 1. Physical hazards which include noise, radiation, electromagnetic fields, ultraviolet radiation and thermal stress, and 2. Chemical hazards which include dusts and fumes, solvent vapors, and gases. Although different methods are employed in measuring physical and chemical hazards, the principles of exposure assessment are similar. In either case both the duration and intensity of exposure is of interest. In construction, where the work environment is perpetually changing, exposure characterization hinges on a strong understanding of the principle tasks performed by each trade.

The need for a task-based approach to assessment of construction health hazards has been described (Susi & Schneider 1995). The importance of using task as the primary building block of an exposure assessment framework is particularly important in construction where repeated performance of trade-specific skills, or specialized tasks, form a thread of continuity in the transient career of a construction worker. Characterizing the life-time exposure profile of the *journeyman* requires a knowledge first and foremost of the principle tasks of their trade. Attempting to characterize exposures by occupation alone is of limited value for the mobile construction workforce and can result in grouping together workers whose exposures may differ by several orders of magnitude. For example, the degree of exposure risk for roofers is dependent on the primary tasks they perform. A survey of a built up roofing operation showed roofers engaged in kettle operation to be exposed to asphalt at levels several orders of magnitude greater than roofers engaged in mopping asphalt and other roofing tasks. These data suggest that kettle operators are at greater risk of disease associated with asphalt fumes and that engineering controls and work practices aimed at reducing emissions from roofing kettles are needed.

As will be described later in this chapter, a framework for characterizing exposures in construction beginning at the fundamental level of task has been developed and used by the Center to Protect Workers' Rights. This framework, a Task-Based Exposure Assessment Model for Construction (T-BEAM), uses standardized definitions and survey materials for collection of exposure data by survey teams made up of occupational hygienists and specially trained workers. Standardized definitions for task, process, exposure and other terms used by T-BEAM are presented in Figure 1.

Moving up a tier from task and trade, the type of projects on which a worker is employed throughout their career will also determine the nature and magnitude of exposure risk. Most workers will work for several employers on many diverse projects in any given year. For example, a pipefitter/welder may be employed in new commercial construction for several months and move onto industrial rehabilitation and maintenance work during subsequent months. The pipefitter employed in commercial construction is likely to encounter welding fumes associated with carbon and mild steels while those who construct or renovate industrial facilities are more likely to encounter metal fumes largely dictated by the industrial process. For example, pulp and paper mills, food and beverage, and semiconductor manufacturing facilities have large amounts of stainless steel metal stock. For the pipefitter, boilermaker, ironworker or sheet metal worker who is involved in fabricating, installing or dismantling such equipment, the constituent metals of stainless steel – chromium and nickel – become a concern. Construction workers engaged in industrial rehabilitation and maintenance projects may also be more likely to work in enclosed work areas which serve to amplify the concentrations of airborne toxins absent effective ventilation.

Project An extensive construction undertaking that is the responsibility of a project manager, general or prime contractor. Typically, a large number of contractors and crafts will be involved. For example, the Rock and Roll Museum Project in Cleveland, OH was a major new construction project involving a large number of sub-contractors and crafts.

Job A smaller construction undertaking which only involves one or a small number of contractors and crafts. Or, an undertaking which is part of a larger project. For example, in 1994 the T-BEAM approach was used to survey a boiler rehabilitation job which was part of a larger power plant "turn-around" project.

Process A series of steps or group of tasks performed for a singular objective. For example, the boiler rehabilitation job described above involved the following major processes: 1) scrapping out or removing old boiler tubes; 2) prepping and milling old tubes to be joined to new tubes; and 3) joining new tubes to old tubes.

Task A specific piece of work that is common to the construction trades. For example, the process of scrapping out old boiler tubes, which is described above, involved the tasks of torch-cutting, rigging and material handling. In construction, there are a number of tasks which are specialized skills. For example the task (or skill) of welding is central to a number of trades including ironworkers, pipefitters, sheet metal workers and boilermakers.

Task Elements The sequential steps required to perform a task. For example, building a set of saw horses would require the following steps: 1) select and gather lumber stock; 2) measure and lay-out cuts; 3) make cuts to measured dimensions; 4) nail together cut pieces.

Task Variables The tools, materials, environmental conditions, work practices and other factors that may impact on either the type or magnitude of exposure hazard during the performance of a task.

Exposure The term used to describe contact with environmental agents that have the potential to cause harm. The level or concentration at which an exposure occurs in combination with the duration of exposure must be assessed to determine the degree of potential hazard.

Figure 1. T-BEAM Definitions.

1.2 *OSHA standards requiring exposure assessment in construction*

The Occupational Safety and Health Act which provided workers in the United States with 'the right to a safe and healthful' workplace is less than 30 years old (OSHA, 1970). Nevertheless, in this relatively short time frame hundreds of hazards encountered in construction are regulated under standards promulgated since Congress enacted the Act in 1970. For most employers and product manufacturers, these standards and avoidance of workers compensation claims and litigation are driving forces for reducing exposure to health hazards in the workplace. OSHA either implicitly or explicitly requires that employers conduct exposure assessment in at least 33 standards. Twenty five of these standards are comprehensive health standards that fall under Subpart Z - Toxic and Hazardous Substances of the Construction Industry Standards found in Title 29 Part 1926. A complete list of standards requiring exposure assessment appears in Figure 2 (CCH Inc. 1998).

Historically, attention to regulations related to safety have exceeded attention given to health hazards in construction. Yet there are many exposures that cause long term health effects that kill or cause disability among construction workers such as asbestos and silicosis. Asbestosis and silicosis are both types of pneumoconiosis (which literally means 'dusty lungs') that plague construction workers. Silicosis, which continues to kill young workers even as we approach the 21st century, is caused by exposure to dust from silica-containing materials extremely common in construction including abrasives, concrete, brick, granite and masonry (MMWR 1998; Linch et al 1998). Such deaths are entirely preventable and indicative of a break-down in' the regulatory framework intended to protect workers.

The mining industry is comparable to construction in many respects and serves as a useful comparison for evaluating silicosis prevention efforts. The mining industry has

29 CFR Subpart D Occupational Health and Environmental Controls

1926.52 Occupational noise exposure
1926.53 Ionizing radiation
1926.54 Nonionizing radiation
1926.55 Gases, vapors, fumes, dusts, and mists (531 regulated agents)
1926.60 Methylenedianiline
1926.62 Lead

29 CFR 1926 Subpart E Personal Protective and Life Saving Equipment

1926.103 Respiratory protection

29 CFR 1926 Subpart Z Toxic and Hazardous Substances

1926.1101 Asbestos
1102 Coal tar putch volatiles; interpretation of term
1103 13 carcinogens (4-Nitrobiphenyl, etc.)
1104 alpha-Naphthylamine
1106 Methyl chloromethyl ether
1107 3,3'-Dichlorobenzidine (and its salts)
1108 bis-Chloromethylether
1109 beta-Naphthylamine
1110 Benzidine
1111 4-Aminodiphenyl
1112 Ethyleneimine
1113 beta-Propiolactone
1114 2-Acetyl-aminofluorene
1115 4-Dimethylaminoazobenzene
1116 N-Nitrosodimethylamine
1117 Vinyl chloride
1118 Inorganic arsenic
1127 Cadmium
1128 Benzene
1129 Coke oven emissions
1144 1,2-dibromo-3-chloropropane
1145 Acrylonitrile
1147 Ethylene oxide
1148 Formaldehyde
1152 Methylene chloride

29 CFR Part 1910 Excerpts From General Industry Standards

1910.1051 1,3-Butadiene

Figure 2. OSHA Standards Requiring Exposure Assessment.

seen a substantial decrease in silicosis in recent decades. Anthracosilicosis, or Black Lung, was a well known killer in the coal fields for much of this century. The United Mine Workers of America (UMWA) were successful in establishing programs that have dramatically reduced the occurrence of this disease among miners. Included among these programs was the establishment of the Black Lung Benefits Program which provides economic relief to victims of Black Lung (or anthraco-silicosis). Revenues for the program come from the mining industry which pays into the fund at a rate commensurate to the tonnage of coal mined. A residual benefit of this system is that the cost of Black Lung is not hidden in the personal lives of miners but represents a sizable expenditure accounted for each year in public Congressional record.

Despite the disabling effects of silica dust, it is only one of 561 agents for which OSHA has assigned a permissible exposure limit (PEL) but has yet to develop a comprehensive standard with provisions for training, medical monitoring and engineering and work practice control requirements. Nevertheless, employers who come within the scope of OSHA regulations must control workplace exposures to those agents assigned a PEL using engineering and administrative controls as the primary means of control. Where such controls are not feasible, personal protective equipment which adequately reduces exposures to below the PEL must be provided to employees as part of a comprehensive respiratory protection program (CCH Inc., 1998). Implicit in the requirement to control health hazards to a designated airborne concentration and to ensure that control measures are effective in reducing exposures to these levels in the need to conduct exposure assessment.

1.3 *Reasons for conducting exposure assessment*

Compliance. There are three principle reasons why exposure assessment occurs. The most common reason in the American workplace is to determine whether or not an employer is providing a working environment in compliance with OSHA standards. As previously discussed, there are 100s of chemical agents for which OSHA has assigned a PEL. In addition, OSHA requires regular air monitoring for a number of substances regulated by comprehensive standards such as lead, cadmium, arsenic and asbestos. Failure to conduct exposure assessment is a common cause for OSHA citations.

Finally, employers have a 'general duty' to protect workers from recognized hazards that can cause serious impairment or death even absent an OSHA standard. For example, OSHA does not have a standard governing heat stress. However, because of the serious and recognized risk that heat stress poses to workers employers have a 'general duty' to protect workers from this hazard. The American Conference of Governmental Industrial Hygienists (ACGIH) provide guidelines for work-rest regimens intended to protect against the adverse effects of heat exposure. The wet-bulb globe temperatures (WBGT) is used to measure exposure and takes into account ambient temperatures, radiant heat and humidity – all factors in human response to heat. Although these guidelines are not legal standards, they are internationally recognized criteria for exposure assessment and control of heat stress. The example of heat stress serves not only to illustrate an application of the OSHA General Duty Clause but also the importance of involving professional occupational hygienists in the design of a comprehensive health hazard assessment and control program. Occupational hygienists have an ethical duty to protect workers from occupational health hazards and should be familiar enough with the literature and practice of industrial hygiene to go beyond minimum legal requirements if necessary.

Health hazard surveillance. Health hazard surveillance is a multi-disciplinary process requiring the involvement of occupational hygienists, epidemiologists, toxicologists and other disciplines knowledgeable in the health outcomes of concern. Depending on the discipline involved, surveillance may take on different meanings. For the epidemiologist, surveillance may be used to describe general trends within a given group. The toxicologist aims to determine with some accuracy the personal exposure levels to environmental contaminants that produce illness or physiological impairment in workers. The occupational hygienist seeks to apply this information in the collection of relevant exposure variables and application of suitably protective controls. The complexity and difficulty in achieving these objectives is compounded by several factors. First, workplace exposures are extremely variable. This is particularly true in industries where exposures are poorly controlled. Secondly, for most chemical agents, the human response is also variable, therefore a comparable dose of contaminant for one worker may elicit very different disease outcomes than in another. Epidemiologists and toxicologists are charged with determining the degree of health risk associated with specific agents. However, their estimates of risk are limited by the quality of exposure data on which their conclusions are based. For most chemical agents there is far from sufficient published data suitable for characterizing the full array of disease outcomes for any given agent.

Ideally, exposure data is collected in such a way that exposure distributions for similarly exposed populations can be accurately characterized. The biggest obstacles to accomplishing this are: 1. 'Worst-case' exposure assessment efforts which rest on the assumption that an occupational hygienist could determine a priori when such exposures are likely to occur and hence prevent an employer from being cited by OSHA for exceedance of a PEL; 2. Insufficient exposure measurements to capture the full range of variability within environmental air monitoring data; and 3. The failure to record descriptive information associated with work processes and environmental conditions necessary to group workers into similarly exposed groups and determine key exposure variables. Rappaport has described in detail the inherent statistical nature of exposure assessment and hence the importance of collecting exposure data in an unbiased manner (Rappaport 1991). Provided such data is representative of the full range of exposures within a population of workers and incorporates sufficient descriptive information with which to analyze variables that effect exposure, an occupational hygienist can design an assessment strategy for both compliance and health surveillance.

Evaluation of control technologies. Perhaps the most important but underutilized purpose for exposure assessment is evaluation of control technologies. Employers have a legal obligation to design work processes to minimize occupational health hazards. Both the occupational health profession and OSHA recognize the principle that employers should follow the 'hierarchy of controls' in controlling workplace hazards. This means using engineering controls such as local exhaust ventilation or substitution of hazardous materials with those that are less hazardous (e.g. use of alternative abrasives in lieu of sandblasting) as the primary means of control. Personal protective equipment (PPE) is at the bottom of the hierarchy to be used only as a last resort to ***supplement*** engineering and work practice controls. There are a number of reasons for this approach, not least of which is that use of PPE fails to control hazards at the source and is therefore the least efficient option. However, regardless of the type of control used, the use of air monitoring to verify that controls are working is essential. In the case of engineering controls, monitoring with direct reading instruments provides a quick and simple method for evaluating the effectiveness of controls when first implemented and as part of a regular maintenance program. For those researchers who have used unbiased approaches to exposure assessment and compared exposure distributions among workers using engineering controls with those not using such equipment, a very important and practical finding has resulted. Not only does use of engineering controls such as ventilation reduce mean exposures among sampled workers, but exposure variance is also significantly smaller. This effect is extremely useful for employers who must ensure that the upper range of employee exposures never exceeds an OSHA PEL.

Although not considered an engineering or work practice control, respiratory protection is often utilized because it is viewed as a relatively inexpensive and simple approach to reducing personal exposure to airborne contaminants. However, use of PPE triggers the need for exposure assessment since personal exposure levels to specific agents must be determined when selecting an adequate respirator and ensuring that pre-assigned respiratory protection factors are sufficient. Professional knowledge and experience may enable an occupational hygienist to make an intelligent *guess* as to what personal exposures are among a group of workers. However, calibrated air monitoring equipment and validated analytical methods are necessary to actually *measure* the concentration of contaminants within the breathing zone of workers.

2 OCCUPATIONAL EXPOSURE LIMITS – BENCHMARKS FOR EVALUATING EXPOSURES IN THE WORKPLACE

Interpretation of exposure data requires use of some benchmark with which to gauge the severity of measured contaminant levels. Occupational exposure limits (OELs) can be broadly classified into two major categories: 1. Legal OELs which in this country are referred to as Permissible Exposure Limits (PELs) and are enforced by OSHA as previously discussed; and 2. Recommended exposure limits.

In the US, the two major sources of recommended OELs are the American Conference of Governmental Industrial Hygienists (ACGIH), which publishes and updates Threshold Limit Values (TLVs), and the National Institute for Occupational Safety and Health, which reviews specific chemical hazards and publishes Recommended Exposure Limits (RELs). RELs serve as health based recommendations to be considered by OSHA during rulemaking for specific agents. Several authors have described the process and limitations of setting OELs (Castleman & Ziem, 1988; Vincent, 1998). However, since worker health can only be protected if exposures are maintained below a level thought to result in adverse health outcome, an understanding of the selection and suitability of OELs is needed to interpret personal exposure measurements.

OSHA PELs. OSHA PELs originate from one of two sources: 1. Adoption of TLVs published between 1968 and 1971; or 2. Establishment of new PELs through the rule-making process for comprehensive health standards. In either case, OSHA standards result from of a legal process which often serves to generate PELs that are weaker than desirable from a public health point of view. One very practical reason for this is that the rulemaking process is slow and the ability to update PELs is extremely difficult resulting in standards that are outdated and fail to reflect current science. Second, OSHA must consider the economic feasibility of controlling exposures to proposed limits in the industries effected by new standards. As a result, the input of those industries weighs in considerably to the rulemaking process. Rarely do such interests support proposed exposure limits which serve to increase the costs of controlling hazards to lower levels. Instead, the prevailing force is towards weakening proposed limits. As a result, OSHA standards, including PELs, often reflect a compromise between occupational disease prevention and the economic interests of those industries required to comply with standards.

ACGIH TLVs and NIOSH RELs. Under the OSH Act, the Department of Health and Human Services, through NI OSH, is charged with reviewing occupational health hazards and recommending exposure limits to OSHA. Dozens of criteria documents have been published by NI OSH which contain Recommended Exposure Limits (RELs) and the methods and rationale for those recommendations. Unfortunately, with tens of thousands of chemical hazards in commerce and the number continuously growing, NI OSH has been unable to keep pace with the ever-growing demand for RELs. Nonetheless, NI OSH criteria documents are an important source of information when selecting an OEL given they are based on the weight of scientific literature available at the time a REL is set.

Threshold Limit Values (TLVs) are consensus limits that have been published by the American Conference of Governmental Industrial Hygienists (ACGIH) since 1946. They are defined as:

> 'the time-weighted average concentration for a conventional 8 hour workday and a 40 hour work week to which it is believed that nearly all workers may be repeatedly exposed day after day without adverse effect' (ACGIH 1998).

Like all OELs, TLVs are designed to protect against a finite set of health outcomes based on interpretation of the current available literature. However, because TLVs are recommended standards they can be and are periodically reviewed and updated. In the past decade, the process by which TLVs are set has come under greater scrutiny which may have served to strengthen the process by which they are established. However, selection of a TLV is still limited by subjective interpretation of the current science which is often far from conclusive.

Regardless of what OELs are selected for comparison against measured exposure levels, review of the basis of the OELs and a critical evaluation as to their suitability given the current literature is a necessary part of a hazard assessment. Given the limitations described above, the professional occupational hygienist and those charged with evaluating and controlling health hazards should take the following steps in interpreting exposure data: 1. Evaluate the adverse health outcomes of concern for selected exposure risks based on current literature; 2. Review OEL health effects criteria including endpoints for which limits were set, literature reviewed and the date established; 3. Select the OEL(s) that best protects against all identified adverse health effects or the most sensitive endpoint; and 4.Design a control strategy aimed at reducing exposures to some fraction of the OEL which takes into account measured variability in exposure.

2.1 *The construction workforce*

The construction workforce is made up of many trades each with specialized skills. In the union sector these skills are cultivated through formal classroom training and on the job work experience completed under the supervision of other journeymen workers. Building trades unions have long served to provide the continuity and training necessary to preserve trade-specific skills through both strong and weak economic markets. The organiza-

International Association of Heat and Frost Insulators and Asbestos Workers (1904)
International Brotherhood of Boilermakers, Iron Shipbuilders, Blacksmiths, Forgers and Helpers (1881)
International Union of Bricklayers and Allied Craftsmen (1865)
United Brotherhood of Carpenters and Joiners of America (1881)
International Brotherhood of Electrical Workers (1891)
International Union of Elevator Constructors (1901)
International Union of Operating Engineers (1886)
International Association of Bridge, Structural and Ornamental Iron Workers (1886)
Laborers' International Union of North America (1903)
Tile, Marble, Terrazzo Finishers, Shopworkers and Granite Cutters International Union (1901)
International Brotherhood of Painters and Allied Trades (1888)
Operative Plasterers and Cement Masons' International Association (1882)
United Union of Roofers, Waterproofers and Allied Workers (1919)
Sheet Metal Workers' International Association (1888)
United Association of Journeyman and Apprentices of the Plumbing and Pipe Fitting Industry (1889)

Figure 3. Labor Organizations Representing Building Trades Workers in the U.S. and Canada.

tion of the construction workforce as we know it today dates back largely to the late nineteenth century. However, the emergence of craft guilds precedes the birth of our nation. Fifteen unions represent the building trades currently numbered at 7.9 million workers in the US (CPWR, 1998). A list of unions representing the US construction work force and the year in which they were chartered appears in Figure 3. Building trades unions are a vital resource for those seeking to characterize occupational exposure in construction since they assist in defining occupations and have a long history of identifying important trade hazards.

Improvement of working conditions, including safety and health, was a driving force in the organization of the construction industry. Reduction of the 'dawn-to-dusk' workday common in the nineteenth century, was among the earliest struggles of the labor movement. The oldest national construction union, the International Union of Bricklayers and Allied Craftsman, founded in 1865, brought together Bricklayer locals from cities such as New York and Baltimore which had began the struggle for the 10 hour day as early as 1823. In asserting their position on the need for a 10 hour work day at a public meeting in 1833, a representative from the Bricklayers said:

> 'We are more exposed, since we start earlier and stop later than any other mechanics and since a man, after 15 hours exposed to the scorching rays of the summer sun, returns to his home worried and dejected, so that he feels no relish for society or improvement, but throws himself down to sleep till the light of the coming day calls him to renew his labors' (BCTD, AFL-CIO 1983).

The relationship between exposure to hazardous agents, the duration of work and the well-being of the worker (as well as the society in which that worker lives) was recognized and articulated by union bricklayers even before the US celebrated its first one hundred years as a nation. Modem OSHA standards and current industrial hygiene reflect this principle in that exposure limits (based on a 40 hour workweek) are reduced when the work week exceeds 40 hours.

2.2 *Use of a task-based exposure assessment model (T-BEAM) for construction*

Common tasks link large number of occupational groups in construction. Since 1993, the Center to Protect Workers' Rights (CPWR) has worked closely with affiliates of the Building and Construction Trades Department, AFL-CIO, in the development and refinement of a Task-Based Exposure Assessment Model (T-BEAM) for construction. T-BEAM uses tasks (or specialized skills) as the central organizing principle for data collection. The defining characteristics of T-BEAM are:

1. Involvement of specially trained journey-level tradespeople, referred to as ' Construction Safety and Health Specialists (CSHSs), in the exposure assessment process; and
2. An emphasis on the identification, implementation and evaluation of engineering and . work practice controls.

The primary objective of T-BEAM is to collect data necessary for occupational disease prevention among construction workers. In the short term, engineering and/or work practice controls are introduced into the workplace and exposure measurements are collected concurrently for both controlled and uncontrolled operations. Such information is intended to assist owners, contractors and unions seeking to incorporate effective hazard control into job design. Direct reading instruments combined with real-time video monitoring supplements traditional air monitoring methods to visually demonstrate the impact controls have on reducing task-related exposure levels.

3 APPLICATION OF T-BEAM TO ASSESSMENT AND CONTROL OF METAL FUMES FROM WELDING AND THERMAL CUTTING

Between 1995-1996, CPWR applied T-BEAM to the collection of welding and thermal cutting fume exposure data among three trades – boilermakers, ironworkers and pipefitters. Collected data was used to estimate exposure distributions among sampled trades and to measure the impact of key determinants of exposure. Data were collected at ten construction sites around the country. Multiple databases were constructed to catalog both quantitative data and descriptive information related to sampled tasks, trades and metal fume exposure levels. Data analysis was structured to answer the following questions:

1. Can exposures be characterized among construction workers absent a fixed work site?
2. Which task variables have significant effects on exposure?
3. What impact does mechanical and local exhaust ventilation have on reducing metal fume exposures among the welding trades?
4. What is the probability of exceeding occupational exposure limits for metal fumes among the sampled trades.

Survey Kits were provided to participants for collection of information in a uniform fashion. Prior to each annual survey campaign, participants attended a 3-4 day training session in survey methodology, use and calibration of air monitoring equipment, and general principles of industrial hygiene.

Each survey was initiated with an opening meeting of local union and contractor representatives, site foreman and stewards for the workers being sampled, and the T-BEAM survey team. During the opening meeting, an initial survey form was used to define the project, job, trades and tasks which were to be included in the scope of each survey.

Selection of workers for whom exposure measurements were taken was based on two criteria: 1. They would be engaged in hot work tasks for at least 60 minutes (cumulative time) during the day selected for sampling, and 2. The work they were performing was considered representative of 'typical' work. The importance of not biasing the collected data by selecting workers who might represent 'worst case' or 'best case' exposure scenarios was emphasized during training and at opening meetings. Survey teams were instructed to sample for at least 7 hours of an 8 hour work day, removing air monitoring equipment from workers only at lunch breaks. NIOSH validated analytical methods were used to determine exposures to both total particulate and selected constituent metals.

Occupational hygienists remained on site for the first 3 days of data collection to ensure their journeyman partners were carrying out exposure monitoring, calculating air volumes and filling out laboratory submittal forms correctly. In addition to total particulate, up to 3 constituent metals were selected by survey teams for laboratory analysis. Manganese was selected where mild or carbon steel was being welded or cut and chromium and nickel was selected when stainless steel was the base metal. In the case of painted steel, survey teams requested analysis for the appropriate metals given the type of coating involved. CSHSs remained on-site for the remainder of the time (generally a total of 10 work days) with OHs available by phone if support was needed.

4 SELECTION OF OELS FOR EVALUATING T-BEAM DATA FROM WELDING AND THERMAL CUTTING

Welding fumes are a complex mixture of chemical hazards. Because of their extremely small size (0.01 - 10 micrometers in diameter) they are able to penetrate the lower airways of the lung and enter the bloodstream thereby posing a greater systemic hazard than larger particles (AIHA, 1984). Sampling strategies aimed at assessing exposure to welding fumes should measure both total fume concentrations as well as constituent metals which will vary in toxicity. NIOSH has found a 40% increase in lung cancer and has observed excess respiratory disease including chronic bronchitis, pneumonia and reduction in pulmonary functions among welders. These effects have been observed even where exposures have been below the OSHA PEL for welding fumes and respirable particulate of 5 mg/m^3 Because of these findings NIOSH believes exposures should be reduced to the lowest feasible level using state-of-the-art engineering controls. In addition, they recommend that agent specific exposure limits be applied to constituent metals (NIOSH 1988).

Gravimetric sampling for total particulate (which measures contaminant mass by subtracting filter pre-sampling weights from post-sampling weights) in combination with additional analysis for specific metals is the sampling methodology typically employed for assessment of welding fume hazard. Particle size selectors such as those used for sampling respirable particulate are not specified by either NIOSH or OSHA for welding fumes. However, investigators have used cascade impactors to determine the size of manual metal arc or shielded metal arc welding fumes and found the mass median aerodynamic diameter of particles to range from 0.2 to 2.0 micrometers in diameter (IARC 1990).

OELs are only as strong as the science on which they are based. The ACGIH TLV for welding fumes, which is the same as the OSHA PEL, was set at 5 mg/m^3 as an 8 hour time-weighted average (TWA) in 1974 (ACOM 1991). The health effects for which this limit was set are metal fume fever, a transient flu-like condition which occurs from exposure to specific metal fumes, and respiratory tract irritation. Zinc fume, which is generated when welding galvanized steel, is a common cause of metal fume fever. A second OEL which could be applied to interpretation of welding fume exposure measurements is the ACGIH TLV for respirable particulate which was lowered to 3 mg/nr^3 in 1994 (ACGIH, 1996). The ACGIH TLV for respirable particulate was set to protect the body's natural pulmonary defense clearance mechanisms which are compromised by excess dust loading (ACGIH, 1996). Such effects are of particular concern with welding fumes which may contain very toxic metals.

Given the toxicity and particle size characteristics of welding fumes, NIOSH's approach to controlling exposures is clearly the most prudent. However, from a practical point of view, some benchmark with which to compare measured exposures to welding fumes is still needed. The ACGIH TLV for respirable particulate of 3 mg/m^3 serves this purpose best since it is slightly lower than the rather outdated welding fume TLV. As such, the respirable particulate TLV was selected as the best available OEL with which to gage T-BEAM results for welding and thermal cutting. However, the limitations to this approach are as follows. First, because the sampling method employed to measure total fume does not distinguish between particle size, larger non-respirable particles may contribute to total measured mass and thereby overestimate risk for respirable particulate. Second, ultrafine particles (smaller than 0.1 micrometers in diameter) are believed to pose a greater health risk than respirable particulate. As previously discussed, the lower size range of welding fumes falls into this ultrafine region of the particle size gradient. The ACGIH currently states that these particles present significant more risk than larger respirable particles however, they also state there is inadequate date available with which to determine an OEL for these particles and that the respirable particulate TLV should therefore not be applied (ACGIH 1996). Therefore, the respirable particulate TLV may not be protective enough when evaluating welding fume exposure. These countervailing limitations to use of the respirable fume TLV as an OEL for welding fume data illustrate the importance of controlling exposures to the lowest feasible limit and recognizing that any current OEL should be used only as a rough upper limit with which to compare and control workplace exposures.

5 SUMMARY OF T-BEAM SURVEY RESULTS FOR 'HOT WORK' TASKS

Between 1995 and 1996 195 full shift samples were collected for total particulate among 63 workers. During the same years, additional analysis for manganese was requested for 136 of these samples collected among 45 workers. A small sub-set of full-shift samples were also analyzed for nickel (27 samples), chromium (24 samples), chromium VI (9), lead (3), cadmium (2), and zinc (4).

Probabilities that a random measurement would exceed an ACGIH TLV were estimated for each trade. The ACGIH TLV for respirable particulate was selected as the OEL with which to compare total particulate results since welding fumes for reasons already discussed. Table 1 indicates the probability of exceeding the ACGIH TLV for respirable particulate was high for all trades (boilermakers, 100%; ironworkers, 71%; and pipefitters, 31%). Probabilities for exceeding the ACGIH TLV for manganese were lower, but still high particularly for boilermakers (72%) and to a lesser extent ironworkers (15%). However, very few samples were analyzed for manganese.

A summary of exposure measurement means relative to other current occupational exposure limits including the OSHA PELs appears in Table 2. The ratio of means to OELs

Table 1. Probability of exceeding OELs for respirable particulate and manganese.

Trade	Total number of samples	Number of samples exceeding the ACGIH TLV[1]	Estimated probability of exceeding the ACGIH TLV[2]
		Total Particulate	
Boilmakers	20	20	100%
Pipefitters	134	42	31%
Ironworkers	41	29	71%
		Manganese	
Boilmakers	18	13	72%
Pipefitters	105	7	7%
Ironworkers	13	2	15%

[1] ACGIH TLV for respirable particulate = 3 mg/m^3, [1] ACGIH TLV for respirable particulate = 0.2 mg/m^3, [2] Calculation for the probability of exceeding the ACGIH TLV: (number of samples greater than the ACGIH TLV/the number of samples) × 100, example: total particulate, ironworkers: (29/41) × 100 = 70.73%.

Table 2. Summary of T-BEAM Exposures means relative to current Occupational Exposure Limits (OELs) by Analyte.

Analyte	Number of samples	Mean (mg/m^3)	Mean/OEL
Total particulate	195	5.00	100% of PEL (respirable – 5 mg/m^3) 167% of TLV (respirable – 3 mg/m^3)
Manganese	136	0.13	3% of PEL (C) (5 mg/m^3) 65% of TLV (0.2 mg/m^3)
Nickel	27	0.05	5% of PEL (1 mg/m^3) 333% of REL (0.015 mg/m^3)
Chromium	24	0.08	8% of PEL (1 mg/m^3) 16% of REL & TLV (0.5 mg/m^3)
Chromium VI	9	0.006	6% of PEL (C)(1 mg/m^3) 60% of TLV (0.01 mg/m^3) 600% of REL (0.001 mg/m^3)

Current Occupational Exposure Limit: Source: ACGIH: Guide to Occupational Exposure Values – 1997, Abbreviations: OSHA PEL: Occupation Safety and Health Administration Permissible Exposure Limits, ACGIH TLV: ACGIH Threshold Limit Value, NIOSH REL: National Institute for Occupational Safety and Health Recommended Exposure Limit, C: Ceiling – the concentration that shall not be exceeded at any time during the work day, The numbers in parentheses represent the respective PEL, TLV, or REL.

which exceed 100 % include total particulate when compared to the ACGIH TLV for respirable particulate; nickel when compared to the NIOSH REL; and chromium VI when compared to both the REL and the TLV. Hexavalent chromium and nickel are associated with hot work on stainless steel, a common metal in industrial process vessels and piping. For example, overexposures to these agents were measured among pipefitters engaged in hot work on stainless steel process vessels being rehabilitated during a pulp and paper mill 'turn-around' project.

Key variables with respect to mean exposures included trade, task, whether work was performed indoors or outdoors, the intermittency of work and use of ventilation. By trade, mean exposure to total particulate was highest among boilermakers followed by ironworkers and then pipefitters. Thermal cutting appeared to generate more total particulate exposure than welding; however, only six personal exposure measurements were associated with thermal cutting versus 148 associated with welding. On average, exposures associated with outdoor work were lower than exposures associated with indoor work, and indoor exposures were significantly lower in 1996 (mean = 2.24 mg/m^3) than in 1995 (mean = 8.75 mg/m^3). This approximate four-fold reduction in sample means reflects, in part, greater use of local exhaust and mechanical ventilation in 1996. Regarding the intermittency of the work. Figure 4 suggests that exposures are of concern even when workers spend less than 25% of the work day engaged in hot work tasks.

Finally, use of mechanical and/or local exhaust ventilation was shown to be effective in reducing exposures through a comparison of means. The sample mean for 1996 total particulate exposure measurements associated with use of mechanical and local exhaust ventilation was 1.79 mg/m^3 (22 samples) versus 5.46 mg/m^3 (57 samples) for natural or no ventilation. A t-test for the difference in the means of exposure measurements collected with use of local/mechanical ventilation versus natural/no ventilation was highly statisti-

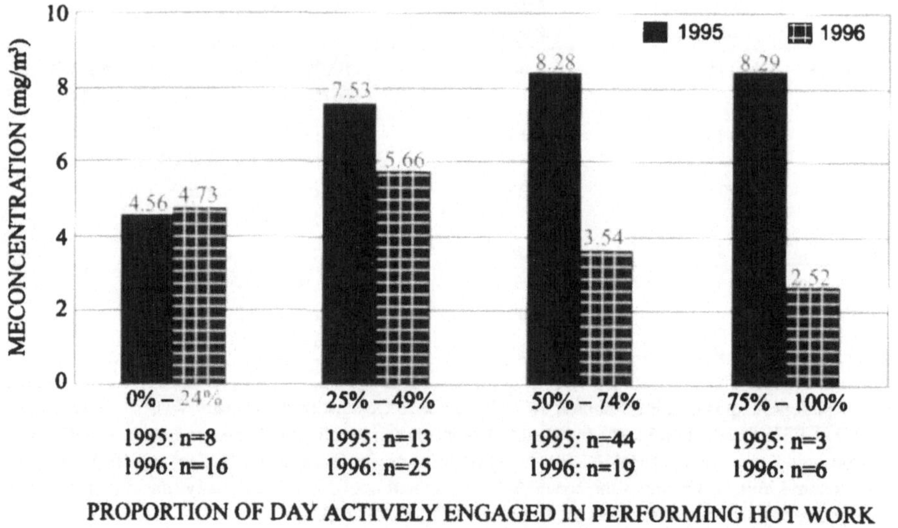

Figure 4. Mean total particulate exposure per shift by proportion of day actively engaged in hot work.

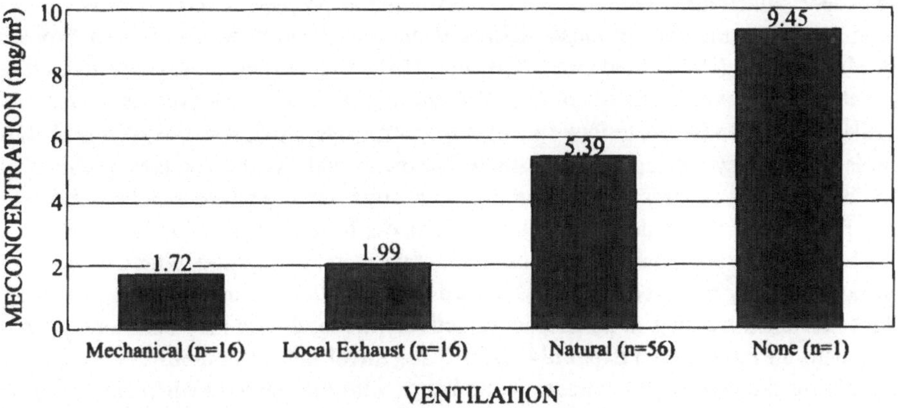

Figure 5. Mean total particulate exposure per shift by ventilation (1996).

cally significant (p-value = 0.0001). Figure 5 illustrates the significant reduction in mean total fume exposure associated with use of ventilation.

This data shows that welding and thermal cutting fumes are likely to pose a significant risk to the health of construction workers. In addition, employers are at risk of exceeding OSHA PELs for welding fumes and other constituent metals dependent on process variables such as ventilation, types of material being welded and the degree of enclosure. Use of mechanical and local exhaust ventilation had a demonstrable impact in reducing exposure risk.

5.1 *Special problems in construction*

As discussed in other chapters (Herrick), the assessment and control of health hazards is a challenging undertaking that requires a multifaceted approach. A methodology and the results for applying a task-based approach have been described. Such an approach has proven to be successful as a starting point in assessing construction health hazards. However, overlaying task-specific hazards is the real-world construction environment composed of multiple trades and employers working under the pressures of demanding work schedules in locations that often are associated with process specific hazards of their own. Given this context, those charged with protecting workers from occupational hazards must consider several problems which make construction uniquely challenging.

Process hazards. New construction and renovation of commercial and residential buildings is only a portion of the construction economy and for some trades a relatively minor share of their work. The contract workforce that maintains and performs capital improvements to industrial facilities is extremely large. Given the seriousness of exposure risks on these projects, any attempt to fully characterize construction health hazards must consider such environments. One of the most important distinctions between the construction employer and the manufacturing employer is that the latter has much more control over the work process. Contractors, on the other hand, are by definition charged with completing a defined scope of work for the client who in turn will direct the contractor on how work will proceed. This phenomenon has particularly important implications in industrial facilities where contractors are engaged for short duration projects involving work on operational process equipment in close vicinity to industrial workers. The OSHA Process Safety Management (PSM) Standard (CCH, Inc. 1998) was written in response to a number of catastrophic explosions which occurred during maintenance of process equipment. The OSHA PSM standard is unique in two important ways. First, it is one of few OSHA standards that go beyond regulating a single hazard to address the complex problem of controlling chemical process hazards through what some may describe as a 'systems' approach. The core of the standard hinges on a 'process hazard analysis' (PHA) that must involve both facility employees as well as contract workers. In addition, both management and workers directly involved in process hazards must participate in conducting the PHA and have sufficient training to participate effectively. Second, this may be the only standard currently in force which explicitly states that general industry employers are legally responsible for the quality of contractor compliance with the standard when employed on the host employers site.

Like any law, the OSHA PSM standard is limited in scope of coverage. However, this standard serves as an excellent model and in some cases a strong impetus for moving up the tier from task-related hazards to process-related hazards. Such an approach would serve the construction industry well in preventing the physical and emotional suffering associated with the routine occurrence of injuries, fatalities and illness that workers continue to endure even in a prosperous economy.

Confined spaces and poorly ventilated work areas. This chapter has concentrated primarily on *chronic* occupational health hazards resulting from exposure to chemical agents. Unfortunately, *acute* exposure to toxic atmospheres account for a large number of deaths among construction workers each year. Often these incidents occur in 'confined spaces' or poorly ventilated work areas. In a review of 109 deaths described in 70 Fatality Assessment and Control Evaluation (FACE) investigations occurring between December 1983 and September 1993, the leading industry with over 18% (n=18) of deaths was construction (US DHHS, 1994). NIOSH defines a confined space as:

> *one which, by design, has limited openings for entry and exit; unfavorable natural ventilation which could contain or produce dangerous air contaminants, and is not intended for continuous employee occupancy.*

Among the 18 construction fatalities investigated by NIOSH, 10 died due to oxygen deficiency, 5 died from toxic atmospheres and 3 died from physical hazards (e.g. trench cave-ins). For many crafts, such as boilermakers, welding in process vessels is common practice. Build-up of toxic vapors and displacement of oxygen can occur quickly with little or no odor warnings in such work settings.

Use of ventilation in confined spaces certainly will reduce exposure risk. But such a reduction may not be sufficient to prevent immediate or latent illness. The amount of contaminant generated and the effectiveness of ventilation at capturing, removing or diluting contaminant will determine the adequacy of ventilation in protecting workers. Given the potentially fatal consequences of working in confined spaces, the effectiveness of ventilation must be assessed in a measurable way. The difficulty of maintaining a healthy atmosphere in confined spaces was illustrated during T-BEAM surveys conducted in 1995. In our survey ofboilermaker exposures to total particulate during a refinery turn-around project, sampled workers welding inside a process vessel had a mean exposure to welding fume 3 times the OSHA PEL and ACGIH TLV even with use of mechanical ventilation.

Safety professionals use the term 'constructibility' to describe a structural design which takes into account not only the load requirements of the finished structure, but also the additional load requirements of the workers, equipment and materials associated with building a structure. Failure to incorporate these considerations into project design can result in accidents of catastrophic proportion. Similar consideration must be given to ventilation needs required to protect the air quality of workers during construction. When construction workers are building non-industrial facilities, sucn as ottice buildings, building ventilation systems are either not yet operational or if operating are not designed to remove the airborne contaminants generated during the construction process. This is particularly problematic in modem 'tight' buildings designed for energy efficiency. Typically, such buildings will rely entirely on heating, ventilation and air conditioning (HVAC) systems for air movement and lack windows that can be opened for temporary natural ventilation. Painters, carpet layers and other trades using solvents in such environments may quickly feel the narcotic effects of solvent vapors and may be at risk of fire and explosions if working in poorly ventilated spaces with flammable materials.

Environmental concerns are yet another driving force for enclosing construction operations without sufficient consideration given to the increased ventilation needs within such enclosures. The best contemporary example of this problem is containment structures used during bridge painting operations designed to prevent release of lead particulate into the ambient air and waterways. Given the enormous concentrations of particulate generated during abrasive blasting operations, the ventilation needed to reduce particulate exposure to permissible levels is difficult to achieve within these structures.

Multi-employer worksites and bystander exposures. As discussed, construction sites are extremely kinetic workplaces where conditions are constantly changing and the work force is always moving. Because trades work side by side, often the worker at greatest risk for exposure to health hazards has nothing to do with the operation generating the hazard. Some have referred to this type of exposure as 'bystander' exposures. Such exposures have been documented in the context a new construction project where a plasterer working approximately 20 feet from a sand-blasting operation had a higher exposure to silica than the crew engaged in sandblasting. The plasterer has no relationship with the contractor engaged in blasting. Such scenarios illustrate the importance of regular health hazard assessments and communication between the multiple employers and trades on site regarding measures for controlling identified hazards. In addition, owners/clients can play an important role in ensuring that contractors are complying with any legal or contractual measures aimed at protecting the construction workforce.

Given the multiple layers and mobility of contractors on any given construction site and the contractual nature of construction, efforts aimed at maintaining a safe and healthful workplace in construction are most likely to succeed if the following are in place:

1. Well defined and enforced job specifications which communicate hazards to contractors and enable them to incorporate hazard prevention into bids; and
2. Joint safety and health committees with employer and employee participation from each contractor and trade on site.

SUMMARY/CONCLUSIONS

There are numerous occupational health hazards in construction that need to be better characterized and controlled. A methodology that measures exposure risk at the task level has been developed and proven effective for evaluating particulate exposures among welding trades. Results indicate a high probability of exceeding OELs for welding fumes among construction workers with the trades at highest risk being boilermakers, ironworkers and pipefitters, in that order. Boilermakers also had a relatively high probability of exceeding the most current OEL for manganese. Other metals of concern are hexavalent chromium and nickel, both of which greatly exceeded NIOSH recommended exposure limits. Local and mechanical ventilation was shown to be an important variable in reducing overall exposures. However, use of mechanical ventilation without quantitative evaluation of effectiveness may not provide adequate protection to workers.

Exposure assessment and control efforts much be part of a more comprehensive safety and health program which emphasizes the use and evaluation of engineering and work practice controls. Because of the diverse environments in which construction workers are employed, exposure assessment strategies should include consideration of process hazards in host facilities and 'bystander' hazards generated by other trades and contractors.

REFERENCES

ACGIH 1998. 1998 TLVs and BEIs – Threshold limit values for chemical substances and physical agents. *American Conference of Governmental Industrial Hygienists, Cincinnati, OH.*

ACGIH 1996. Documentation of the Threshold Limit Values and Biological Exposure Indices. Sixth Edition. *American Conference of Governmental Industrial Hygienists, Inc Cincinnati, OH.* Supplement: Particulate (Insoluble) pp. 1-5.

ACGIH 1991. Documentation of the Threshold Limit Values and Biological Exposure Indices. Sixth Edition.' American Conference of Governmental Industrial Hygienists, Inc. Cincinnati, OH. pp. 1726-1727.

AIHA (1984) *Arc welding and your health: a handbook of health information for welding.* American Industrial Hygiene Association, Fairfax, VA pp. 6-7.

Building and Construction Trades Dept., AFL-CIO 1983. The Builders: The Seventy-five Year History of the Building and Construction Trades Department, AFL-CIO, Washington, D.C.

Beaumont, J.J. & Weiss, N.S. 1980 Mortality of welders, shipfitters and other metal trades workers in boilermakers local no. 104, AFL-CIO. *Am. J. Epidmiol.* 112: 775-786.

Castleman, B.I & Ziem, G.E. 1988 'Corporate influence on threshold limit values.' *Am. J. Ind. Med.*, 1988, (13): 531-559.

CPWR 1998 *The construction chart book.* Center to Protect Workers' Rights. Washington, D.C..

International Agency for Research on Cancer. (1990) *IARC Monographs on the Evaluation of Carcinogenic Risks to Humans: Chromium, Nickel and Welding.* IARC, Lyon, France.

Linch, K.D., Miller, E.W., Althouse, R.B., Groce, D. & Hale, J.M. 1998 Surveillance of respirable crystalline silica dust using OSHA compliance data 1979-1995. *American Journal of Industrial Medicine* 34:547-558

MMWR 1998 Silicosis deaths among young adults – United States, 1968-1994. *Morbidity and Mortality Weekly Review.* May 1, 1998. Vol. 47/No. 16.

NIOSH 1988 *Criteria for a recommended standard: Welding, brazing and thermal cutting.* US Department of Health and Human Services, National Institute for Occupational Safety and Health. Cincinnati, OH. Publication No. 88-110. US Government Printing Office. Washington, D.C.

NIOSH 1998 *Criteria for a recommended standard: Occupational noise exposure.* US Department of Health and Human Services, National Institute for Occupational Safety and Health. Cincinnati, OH.

Occupational Safety and Health Act, Pub. Law 91-596, Section 2193. 91st Congress, Dec. 29, 1970; as amended. Pub. Law 101-552, Section 3101, Nov. 5, 1990.

Occupational Safety and Health Standards for the Construction Industry (29 CFR Part 1926) Section 55: Gases, vapors, fumes, dusts and mists. CCH Inc. Chicago, IL (1998).

Occupational Safety and Health Standards for the Construction Industry (29 CFR Part 1926) Section 64: Process safety management of highly hazardous chemicals. CCH Inc. Chicago, IL (1998).

Occupational Safety and Health Standards for the Construction Industry (29 CFR Part 1926) Section 103: Respiratory Protection. CCH Inc. Chicago, IL (1998).

Rappaport, S.M. & Smith, T.J. 1991 *Exposure Assessment for Epidemiology and Hazard Control.* Chelsea, Michigan, USA, Lewis Publishers, Inc.

Schneider, S. and Susi, P. 1993 Final Report: An Investigation of Health Hazards on a new construction project. Center to Protect Workers' Rights. Washington, D.C.

Susi, P. & Schneider, S. 1995 Database needs for a task-based exposure assessment model for construction. *Appl. Occup. Environ. Hyg.* 10(4): 394-399.

US DHHS 1994 *Worker deaths in confined spaces: a summary of NIOSH surveillance and investigative findings.* National Institute for Occupational Safety and Health. DHHS (NIOSH) 94-103. Cincinnati, OH. p. 20-21.

Vincent, J.H. 1998 International occupational exposure standards: a review and commentary. *Amind. Hygiene Assoc. J.* (59): 729-742.

CHAPTER 7

Overview of health hazards in construction

ROBERT F. HERRICK
Harvard School of Public Health, Boston

ABSTRACT: Health hazards in construction can be recognized, evaluated, and controlled, provided we recognize the unique aspects of construction. The methods which have been used to study and control hazardous exposures in general industry provide a start, but they may not be suited to the complex, constantly changing conditions on a construction site. It is possible, however, to identify a set of exposures that result from work activities commonly performed by one or more of the trades involved in construction. The result of this approach is a comprehensive assessment of health hazards associated with construction work.

1 INTRODUCTION

> *'There are several important approaches to the subject of safeguarding health in construction. One is that of the Safety Engineer who may limit his interpretation of health protection to the prevention of injury by accident; another is that of the industrial hygiene engineer who believes that health conservation may be influenced by almost everything under the sun. For this reason the concept of the latter is apt to be broader than that of the Safety Engineer.'*
>
> (Lippmann 1997)

In approaching a topic such as an overview of health hazards in construction, an honest author will begin by acknowledging at least two limitations. The first is that the overview will be an incomplete summary, and second, that this is not the first time such a review has been attempted. The truth of the first acknowledgment will be evident to the readers of this chapter. For the second limitation, we can refer back to 1942 when a very comprehensive review of health hazards in construction was presented to an audience of civil engineers. The author, Marion Trice, alerted his audience to the hazards of silica, carbon monoxide, oxides of nitrogen, metal fumes, concrete and cement dust, coal tar, oil and grease, lack of safe drinking water, use of gasoline to clean the skin, and heat stress, among other hazards. Trice closed by reminding them that '...the supervision of all kinds of construction is the task of a civil engineer; and the man in charge should be able to recognize potential health hazards and know when to provide the necessary safeguards.' (Lippmann 1997) The fact that nearly all these hazards are still present on today's construction sites is not in any way a criticism of Trice's message – if anything it validates his advice that the recognition and control of health hazards must be an ongoing part of the planning and supervision of all construction activity.

This discussion will build upon the foundation we have from Trice and others who have preceded us in the effort to safeguard construction work by reducing health hazards. Despite the complexity of the working conditions in construction, we can employ an approach which has been proven effective in other settings: the prevention of workplace disease through the recognition, evaluation and control of health hazards. This model of disease prevention is the foundation of industrial hygiene practice, and it can be applied to reduce occupational hazards in construction.

2 CLASSIFICATION OF HAZARDS

Many of the health hazards that construction workers encounter are not unique to the construction trades. What sets the construction worker apart from the production worker who manufactured the bricks, chemicals, planks, asphalt and other materials of construction is the way in construction workers contact these materials. Most of these materials are manufactured in well-controlled facilities where the chance of significant exposure is relatively low. However, after being produced in settings where industrial hygiene and safety systems are in place to reduce hazards, these materials may then used by construction trade workers in uncontrolled settings, where the result may be dramatically higher exposures than were found in the production facility. An example is the contrast in exposure between workers who manufacture paints, and workers who apply these same materials. While the production worker is very likely to be protected by closed manufacturing systems which contain solvent vapors and hazardous pigments, the construction painter may apply this same material by spraying it into the air. While this is only one example, it illustrates the extreme differences between the exposure situations, and resulting health hazards which construction trade workers face compared to workers in general industry.

Recognizing these differences in the extent of exposure and magnitude of health risks to construction workers, we can still apply the industrial hygiene approach of recognition, evaluation and control of health hazards on the construction site. This process starts with the recognition of potential health hazards, and the evaluation of the type, and magnitude of the risk which may result from contact with the hazard.

For purposes of hazard control and disease prevention, contaminants are classified largely on the basis of their physical and chemical characteristics, as these characteristics are important determinants of the route of exposure, and the resulting toxicity. Workers may be exposed to contaminants by inhalation, by absorption through the skin, by ingestion or by injection (e.g., through accidental puncture wounds). Inhalation and skin absorption are the primary routes of exposure for most materials in the occupational environment. In cases where poor hygiene practices such as consumption of food and beverages in the contaminated work areas are allowed, ingestion may an important source of exposure.

3 HAZARDS AND RELATED HEALTH EFFECTS

Environmental agents are broadly classified as either physical hazards or health hazards:

Physical Hazards: Materials such explosives, flammable or combustible liquids, oxidizers, compressed gases, organic peroxides, pyrophoric materials, unstable (reactive) chemical or water reactive chemicals are regarded as physical hazards by the Occupational Safety and Health Administration (OSHA) in its hazard communication standard. Other exposures in the workplace such as excessive noise, ionizing and non-ionizing radiation and temperature extremes are other examples of physical hazards. Ergonomic factors are a broad set of conditions and work situations which are increasingly important causes of injury in the workplace. Repetitive motions, often in awkward positions, are responsible for a number of chronic trauma disorders such as Carpal Tunnel Syndrome.

Health Hazards: Health hazards include chemicals and biological materials capable of producing adverse acute or chronic health effects. Chemical hazards include exposures to chemical mists, vapors, gases or particulates (dusts and fumes) through inhalation or by absorption through the skin. OSHA defines hazardous chemicals to include carcinogens, toxic or highly toxic agents, reproductive toxins, irritants, corrosives, sensitizers, hepatoxins, nephrotoxins, agents which act on the hematopoietic system and agents which damage the lungs, skin, eyes, or mucous membranes. Biological hazards include exposures to infectious or immunologically active agents such as molds, fungi and bacteria.

This discussion will focus on construction health hazards as defined above, however the significance of risks posed by the physical hazards should not be minimized. The losses resulting from excessive noise, vibration, and musculoskeletal stresses are very substantial in construction trades.

3.1 *Health effects related to construction work*

There is a substantial amount of data that documents the hazardous nature of construction work. Information from the Bureau of Labor Statistics (BLS) for example, shows that construction workers have higher rates of work-related death, injury and illness than workers in the manufacturing or service sectors. Of these three adverse effects (death, injury and illness), work-related illness is by far the most likely to be under reported. When a fatality or injury occurs on the job, there is usually a reasonably clear relationship between the cause and the effect. This relationship is much less clear in the case of work-related illness, however. Many of the illnesses which result from contact with workplace exposures appear months or years after the initial exposure, so the relationship between cause and effect is obscured. In other cases, the health effect or disease, such as lung cancer, may be caused by any number of factors, some of which are associated with work, others of which are not. The result is that work-related illnesses among construction workers are substantially under reported.

The major categories of health effects associated with exposures in construction can be summarized as follows: sensory irritation, physical irritation, respiratory effects, narcosis, neuropathic effects, cancer, general systemic effects, ocular effects, dermal effects, cardiovascular effects, and liver/kidney effects. Each of these categories is discussed, with examples of substances commonly used in construction which are associated with each type of effect. This listing of health effects, with examples of construction site exposures that have been associated with them, is not intended to suggest that these exposures will inevitably cause these effects, or that every worker with the exposure will develop the illness or conditions identified. There is a wide range in the nature and magnitude of response between people. In addition, there is wide variability in environmental factors including the concentration, stability and physicochemical properties of the agent in the exposure environment and the duration, frequency and route of exposure. In addition, acute and chronic exposures to a chemical may result in different pathological manifestation. The following sections discuss the broad types of responses which may occur following exposure to hazards in the construction environment.

3.2 *Sensory irritation*

Exposure to many substances is associated with the development of sensory irritation. The irritant response is initiated when these substances come into contact with the eyes, mucous membranes (including those in the respiratory tract) or the skin. The symptoms of sensory irritation include stinging, itching, or burning of the eyes; tearing or lachrimation, a burning sensation in the throat or nasal passages, rhinitis (nasal irritation), cough, sputum production, chest pain, wheezing, and dyspnea (difficulty breathing). In most cases, the onset of symptoms occurs immediately after exposure to the irritant, therefore, there is a clear association between the exposure and the effect. The effect of sensory irritants may cause severe discomfort and may be seriously disabling. For example in the case of dyspnea or wheezing, the tearing, pain, and irritation resulting from exposure can be severe, and in some cases may be disabling. In addition to the primary effects of irritants, workers exposed may be distracted and more likely to have accidents which endanger themselves and other workers.

The eye irritation resulting from exposure to irritants is caused by the stimulation of the sensory nerve endings in the cornea. The underlying mechanism by which the irritant response is produced is not well understood. The symptoms of eye irritation are usually transient and do not persist after cessation of exposure. However, exposure to high levels of lachrimators may produce corneal or conjunctival injury which requires medical treatment.

Sensory irritation of the respiratory tract primarily affects the upper regions and produces an increase in sputum production. Effects include inflammation of the nasal passages, trachea, and upper bronchial tree, and decreased ciliary clearance. These effects are apparent as a burning sensation in the throat and nasal passages, coughing, sneezing, and acute bronchitis. Wheezing may also be apparent, especially in people with a prior history of airway disease. In cases of extreme exposure, there may be chemical burns of the tissue in the lower airways and alveoli. The result may be a rapid collection of fluids in these regions (pulmonary edema) which can cause hypoxia and difficulty breathing. Irritants produce a pattern of generalized, non-specific tissue inflammation, and destruction may result at the area of contaminant contact. Some irritants produce no systemic effect because the irritant response is much greater than any systemic effect, while some also have significant systemic effects following absorption – for example, hydrogen sulfide absorbed via the lungs.

At high concentrations, irritants may cause a burning sensation in the nose and throat (and usually also in the eyes), pain in the chest and coughing producing inflammation of the mucosa (tracheitis, bronchitis). Examples of irritants are gases such as chlorine, fluorine, sulfur dioxide, phosgene and oxides of nitrogen; mists of acids or alkali; fumes of cadmium; dusts of zinc chloride and vanadium pentoxide. High concentrations of chemical irritants may also penetrate deep into the lungs and cause lung edema (the alveoli are filled with liquid) or inflammation (chemical pneumonitis).

Highly elevated concentrations of dusts which have no chemical irritative properties can also mechanically irritate bronchi and, after entering the gastrointestinal tract, may also contribute to stomach and colon cancer.

Exposure to irritants may result in death if critical organs are severely damaged. On the other hand, the damage may be reversible, or it may result in permanent loss of some degree of function, such as impaired gas-exchange capacity.

3.2.1 *Examples found in construction*

A large number of chemicals commonly found in construction have irritant effects (Table 1). Examples are acetone, ammonia, chlorine, copper fume, ethylene glycol (antifreeze), pyrolysis products from rosin core solder (formaldehyde), sodium hydroxide (lye), sulfuric acid, turpentine, VM&P naphtha, and xylene.

Table 1. Sensory irritants.

Substance	Total exposed FTE (person years)	FTE exposed above PEL (person years)
Ethylene glycol	314657	7950
Xylene	376742	4357
Acetone	94130	670
VM&P Naphtha	24221	473
Ammonia	66969	316
Isopropyl alcohol	137103	200
Chlorine	6601	54

3.3 *Physical irritation*

There are a number of substances found in construction activities that are capable of eliciting a set of adverse effects because of their irritant properties (Table 2). These substances differ from the sensory irritants in that they do not fit well into the classification systems based upon a specific target organ of irritation. Many of these substances are airborne particles and dusts which are believed to be of generally low toxicity, but they can still have adverse effects because of the general physical irritation that results for exposure at high concentrations. These materials (usually dusts) are deposited in the eyes, ears, nasal passages, upper respiratory tract, and on the surface of the skin. The result of these exposures may be that workers have difficulty seeing, may cough uncontrollably, or develop conjunctivitis or dermatitis. The biological effects of exposure to these dusts are long-term in nature, and they may overload the body's protective mechanisms, for example, leading to respiratory diseases.

3.3.1 *Examples found in construction*

Physical irritants common in construction are aluminum metal dust, calcium carbonate (limestone), cellulose dust, gypsum and kaolin dust, marble, perlite, plaster of Paris, Portland cement, and starch.

Table 2. Physical irritants.

Substance	Total exposed FTE (person years)	FTE exposed above PEL (person years)
Titanium dioxide	222499	5355
Portland cement	409291	3905
Molybdenum	4974	1696
TDI	20563	401

3.4 *Respiratory effects*

A large of number of materials used in construction may present hazards to the respiratory system. These substances cause acute pulmonary edema, alveolar damage, and chronic respiratory disease through mechanisms of cellular damage or fibrosis. These materials can produce health conditions that are permanent, disabling and life threatening. The diseases resulting from exposure to these materials may be acute or chronic. The acute pulmonary diseases may be life threatening, while the chronic disease resulting form long-term exposures may progress slowly starting with only a few, non-specific signs. Some of the respiratory conditions are irreversible, and may continue their progression even after exposure to the harmful agent has ceased. People who have preexisting respiratory conditions including asthma, bronchitis, emphysema, and pulmonary fibrosis are at extra risk from exposure to these substances. In additions, tobacco smoke is known to cause, or at least aggravate all the pulmonary conditions of concern.

There are three general classes of contaminants which can affect the respiratory system: aerosols, gases, and vapors. Each is briefly discussed, with reference to the characteristics which are important in determining the respiratory hazard associated with human exposure.

Aerosols: Aerosols are composed of liquid droplets or solid particles of fine enough particle size to remain dispersed in air for a prolonged period of time. Aerosols may also be referred to as airborne particulate matter. The hazard associated with airborne particulate matter is a function of: 1. The biological activity of the material, 2. Concentration of the airborne material, and 3. Airborne particle size. Particle size effects site of deposition within the respiratory system. Many occupational diseases such as silicosis and asbestosis are associated with material deposited in different regions of the respiratory tract. Criteria have been developed to define critical size-fractions most closely associated with various health effects and are defined as follows:

- *Inspirable fraction:* This is the fraction of airborne particulate matter which can be deposited any place with the respiratory tract.
- *Thoracic fraction:* Those particles which are hazardous when deposited anywhere within the lung airways and the gas-exchange region.
- *Respirable fraction:* Those particles which are a hazard when deposited in the gas-exchange region of the lungs.

Gases and vapors: Gases are formless fluids that expand to occupy the space or enclosure in which they are confined. Materials are generally considered gases if they are predominantly in the gaseous state at temperatures and pressures which would normally be present in the ambient or occupational environments. Vapors are the gaseous form of substances that are normally in the solid or liquid state at room temperature and pressure. Liquids change into the vapor state by the process of evaporation, and mix with the surrounding atmosphere. Many solvents will volatilize to form vapors at normal temperatures and pressures. In addition, application of heat in many industrial operations results in increased vaporization.

Respiratory conditions may be classified into two general types: conditions resulting from damage to the cells lining the airways; and conditions caused by the production of fibrosis. Common conditions resulting from cellular damage are edema and emphysema. When the cells in the airway are damaged, the result is increased permeability of the cell membrane which causes edema, hemorrhage, and subsequent cell death producing areas of dead cells (localized necrosis). Depend upon the nature of the contaminant, the area of cell damage may be lower in the respiratory tract, at the alveolar septa. The result of this type of damage is emphysema. Emphysema is caused by the gradual destruction of the cells in the alveolar septa, resulting in the loss of elasticity in the lung. The impairment resulting from these conditions is not recognizable until it has reached an advanced stage, at which point is it irreversible. Emphysema is caused by ozone, sulfurtetrafluoride, and nitrogen dioxide, among other substances.

Some chemicals cause localized cellular damage which results in edema, the release of fluid from the cells to the open spaces of the airways. Edema may develop gradually, without warning symptoms, so an individual can sustain a life-threatening or fatal exposure without being aware of it. Example of substances which can induce edema are ozone, nitrogen dioxide, and paraquat.

A number of material commonly found in construction can result in the development of a group of chronic lung disorders termed pneumoconioses (presence of foreign substances in the lung). The term pneumoconiosis includes many fibrotic conditions of the lung (diseases characterized by scar formation in the interstitial connective tissue). Pneumoconioses are due to the inhalation and subsequent selective retention of certain dusts in the alveoli, from which they are subject to interstitial sequestration. Pneumoconioses are characterized by specific fibrotic lesions, which differ in type and pattern according to the dust involved. For example, silicosis, due to the deposition of crystalline-free silica, is characterized by a nodular type of fibrosis, while a diffuse fibrosis is found in asbestosis, due to asbestos-fiber exposure. Certain dusts, such as iron oxide, produce only altered radiology (siderosis) with no functional impairment, while the effects of others range from a minimal disability to death (Lippmann 1997).

3.4.1 *Examples found in construction*

Examples of substances found in construction which cause respiratory effects include asphalt fumes, carbon black, fibrous glass, iron oxides, dust and fumes (from welding and abrasive blasting), mineral wool fiber, nitrogen dioxide (from engine exhaust), ozone, silica, and wood dust (Table 3).

Table 3. Respiratory irritants.

Substance	Total exposed FTE (person years)	FTE exposed above PEL (person years)
Silica quartz (crystalline)	852083	30583
Silica cristobalite	68938	11783
Nickel, soluble	20137	2139
Asphalt fume	300786	1643
Fibrous glass	180484	498
Wood dust (hard wood)	152322	304
Wood dust (soft wood)	395217	189

3.5 *Narcosis*

Narcosis is a result of a general depression of the central nervous system. When the CNS becomes significantly depressed, the awareness or consciousness of the exposed person is diminished. Symptoms resulting from CNS toxins include drowsiness, difficulty in concentration, and mood changes. As narcosis progresses, symptoms including slurred speech, dizziness, loss of coordination, loss of consciousness, coma and death. In the great majority if cases, CNS depression is reversible if exposure to the toxic material ceases. Because even the early symptoms of narcosis affect the concentration and coordination of construction workers, these exposures are a very substantial hazard on the construction site. Workers, and their coworkers are at increased risk of injuries and accidents caused by slowed reaction times, incoordination, mistakes and errors in judgement. CNS toxins produce a narcotic effect by a mechanism similar to that of general anesthetics, which produce a reversible effect on electrically excitable neuronal membranes in the central nervous system.

3.5.1 *Examples found in construction*

A number of commonly used construction materials contain compounds capable of producing narcotic effects. Examples are butane, ethanol, gasoline, hexane, naphtha (coal tar), stoddard solvent, styrene, toluene, and trichloroethylene (Table 4).

Table 4. Compounds producing narcosis.

Substance	Total exposed FTE (person years)	FTE exposed above PEL (person years)
Toluene	314940	2291
methyl chloroform (1,1,1,trichloroethane)	112119	1431
n-heptane	34926	85
Trichloroethylene	67359	68

3.6 *Neuropathic effects*

Many chemicals found in construction have been shown to cause severe, and frequently irreversible neurological effects in exposed workers. Neurotoxic agents can affect both the central nervous system (CNS) and the peripheral nervous system (PNS). The CNS is generally considered to be comprised of the brain and spinal cord, while the PNS is the network of nerves that extends throughout the body, connected to the CNS. Chemicals which have a neurotoxic effect are not easy to identify, in part because they may not affect a specific target site. For example, some neurotoxic compounds affect the CNS, but their effects are manifested through changes in the PNS. An example is organic mercury poisoning, which affects the CNS by causing tremor in the limbs. Nervous system toxins can affect motor function, sensory function, integrative processes, and they may cause behavioral changes in exposed persons. Some neurtoxins cause demyelination or damage the neurons, resulting in motor dysfunction such as loss of muscle strength or unsteadiness of gait. Other neurotoxins cause alterations of sensory function such as alterations in touch, pain, temperature sensation, or damage to sight or hearing. Effects in the central nervous system are believed to be caused by neurochemical alterations in the brain. Examples of CNS effects associated with exposure to neurotoxic chemicals are seizures, parkinsonism, intellectual impairment, dementia, cranial neuropathy, and visual disturbances. PNS effects include muscle weakness, which can progress to loss of motor function, and paralysis.

3.6.1 *Examples found in construction*

Chemical which have been associated with neuropathic effects are lead, n-butyl alcohol, n-hexane, Lindane, manganese dust and fume, mercury (both organic and inorganic forms), and methyl bromide (used as a fumigant) (Table 5).

Table 5. Compounds producing neuropathic effects.

Substance	Total exposed FTE (person years)	FTE exposed above PEL (person years)
Petroleum distillates	473443	16387
Lead	105050	6266
n-hexane	63936	2397
mercury (aryl and inorganic)	93184	1471

3.7 *Cancer*

Cancer is a general term for a set of insidious and life threatening diseases which are characterized by the invasion of organ systems by uncontrolled tissue growth. The development of cancer is due to the growth of abnormal tissue comprised of cells that have been altered in such a way as to cause uncontrolled cell growth. As this growth progresses, it interferes with the normal vital functions of the target organ (s). The sequence of events by which cancer is initiated and develops is a complex and multistage process. While the process is not fully understood, it is clear that a number of chemicals and agents found in the construction environment are associated with increased risk of cancer in workers. Some substances are probably associated with the initiation of cancer which occurs when chemicals interact either directly or indirectly with DNA to cause a heritable mutation. These changes in the structure of DNA can cause an alteration in the reading of the DNA sequence during cell replication, producing more altered cells, and eventually the formation of a tumor. The second stage of cancer formation is promotion, which refers to a series of events which facilitate the unrestricted multiplication of initiated cells. this can lead to the progression of disease resulting in the eventual formation of a tumor. In addition, some agents are thought to be complete carcinogens, meaning that they are capable of both initiation and promotion. Exposure to these substances alone is capable of producing cancer.

Evidence of association between exposure to chemicals (and other agents) and the excess risk of cancer is hard to obtain, and frequently surrounded with uncertainty and controversy. One reason do to this is the difficulty in attempting to relate exposure to a specific agent to cancer development in humans is the long latent period, typically from 15 to 40 years, between onset of exposure and disease manifestation. (Lippmann 1997). Another complicating factor is that the disease endpoints (the various types of cancer) are rarely unique to the occupational exposure which is suspects of causing, or contributing to them. The best example is cancer of the respiratory system and cigarette smoking, both independently and in combination with other risk factors such as occupational exposures. The range of diseases classified as respiratory tract cancers are not uniquely associated with specific exposures (with some rare exceptions), so the attribution of a cause and effects association between cancer and a particular exposure is very difficult to establish, given that most workers (and people in general) are exposed to a range of possible cancer risks over a working lifetime.

3.7.1 *Examples found in construction*

There are a number of chemicals and other substances found in construction which have been shown to be associated with increased risk of cancer (Table 6). These include acrylamide, arsenic, asbestos, benzene, beryllium, carbon tetrachloride, chromates, coal tar pitch, formaldehyde, perchloroethylene (also known as tetrachloroethylene, or dry cleaning solvent), and zinc chromates (as found in primers).

Table 6. Substances linked to cancer.

Substance	Total exposed FTE (person years)	FTE exposed above PEL (person years)
Chromic acid/chromates	7764	591
2-nitropropane	1022	23

3.8 *Systemic effects*

Many occupational exposures produce a generalized systemic disease due to their effects upon multiple target sites. These agents can either affect several organ systems simultaneously, or cause a variety of nonspecific adverse effects that are indicative of general toxicity. All of these agents are biologically active, and have been demonstrated to interfere significantly with normal biological processes and organ function. The indicators of this toxicity are usually nonspecific symptoms such as dizziness, respiratory irritation, hematuria, chest tightness, weight loss or gain, lethargy, loss of appetite, nervousness, or gastrointestinal symptoms. An example is a chemical which enters the body through the respiratory system and passes into the bloodstream without any damage to the lungs. When the agent is distributed by the blood circulation to various organs, it can damage them or cause general poisoning and have systemic effects. An example is fumes from metals (such as zinc) which are associated with an acute systemic syndrome known as metal fume fever.

3.8.1 *Examples found in construction*

Substances found in construction that are associated with general systemic toxicity include aluminum welding fumes (respiratory effects), organic arsenic compounds (hemolysis, renal failure), 2-butoxyethanol (also known as butyl cellosolve, associated with sensory irritation), Captan (recurrent uticaria), carbaryl (cholinesterase inhibition), cyanide compounds (headache, CNS effects), Dieldrin (convulsions, headaches, nausea), inorganic lead (CNS damage, reproductive effects, urinary tract damage, blood effects), nickel (metal and insoluble salts, dermal sensitization, cancer, kidney damage), tin (irritation, anemia, gastrointestinal symptoms), and welding fumes (metal fume fever, gastritis) (Table 7).

Table 7. Substances producing systematic effects.

Substance	Total exposed FTE (person years)	FTE exposed above PEL (person years)
Welding fume	106735	11086
Butoxyethanol	64368	1299
Zinc oxide fume	15322	235

3.9 *Ocular effects*

Several chemicals are specific hazards because of their ocular effects (effect on the eye). In most cases, these chemicals have other effects (such as systemic toxicity or irritation), but their ocular effects occur at lower levels of exposure than do the other effects. The damage these chemical cause to the eye can occur in the form of corneal, lens, retinal, ganglion cell layer, or optic nerve damage. These effects may be transient, temporarily disabling, or permanently blinding. Of the corneal effects, changes which result in the loss of transparency of the cornea are the most significant risk. The vapors of some chemicals can cause damage to the corneal epithelium, which may result in burning, blurred vision, tearing, photophobia, scarring, opacity, and loss of vision. In some cases, exposure causes edema of the corneal epithelium, resulting in the production of visual halos(halo vision). N-ethylmorpholine, a chemical used in the production of polyurethane foams has been specifically associated with halo vision.

Retinal effects are generally not the result of exposures to the outer surfaces of the eye. Exposure to chemicals such as methyl alcohol can have a systemic effects, however, as these substances can reach the retina through the blood, and result in retinal changes including edema and hemorrhage, narrowing of the retinal arteries, damage to the optic nerve and loss of vision.

3.9.1 *Examples found in construction*

Chemicals associated with ocular effects include N-ethylmorpholine (used in polyurethanes), hydrogen sulfide, methyl alcohol, and naphthalene (Table 8).

Table 8. Substances producing ocular effects.

Substance	Total exposed FTE (person years)	FTE exposed above PEL (person years)
methyl alcohol	64716	68
hydrogen sulfide	5801	54

3.10 *Dermal effects*

Skin disorders are the most frequently reported occupational disease. An occupational skin injury may be an immediate effect resulting from instantaneous trauma, or brief exposure to a toxic material. Occupational skin diseases result from longer periods of exposure, and may have a longer interval between the time of exposure and the development of symptoms and disease. Skin conditions associated with workplace exposures and conditions can be classified into several clinical patterns. These include contact dermatitis, infections, pilosebaceous unit abnormalities, pigment disorders, and neoplasms.

Contact dermatitis is the most common occupational skin condition. It may be produced by irritants, or allergic contact sensitizers. Contact dermatitis is characterized by reaction at the site of contact with the irritating or sensitizing agent, and redness, swelling, blister formation, exudation, crusting and scaling are frequently observed. Primary contact dermatitis is a reaction of skin caused by exposure to irritating substances. Most (up to 80%) of occupational contact reactions are of the irritant type. Allergic contact dermatitis is distinguished from primary irritant type by a delayed hypersensitivity reaction. It results when sensitized people come into contact with allergens. The most common examples of allergic contact dermatitis reactions are to poison ivy, oak or sumac. Symptoms are similar to those associated with primary irritant contact dermatitis, however, they may be delayed in appearing, so lesions do not become apparent until 2 or 3 days after exposure.

Infections of the skin may result from contact with bacterial, fungal and viral agents. Although these conditions are not unique to construction work, many of these infections are opportunistic and the minor injuries (cuts and nicks) common to construction can place workers at risk of acquiring these infections. Pigment disorders can include stains from materials adhering to the outer layer of the skin (the stratum corneum), or altered myelin pigmentation. Construction workers are at increased risk because of their exposure to large amounts of ultraviolet radiation from sunlight, which in conjunction with photosensitizers such as coal tar pitch can result in enhanced erythema. Neoplasms of the skin including squamous cell carcinoma and basal cell carcinoma can result from chronic exposure to ultraviolet light. This effect is also enhanced by exposure to the polycyclic aromatic compounds found in coal tar pitch.

3.10.1 *Examples found in construction*

The following materials found in construction have been associated with occupational dermatoses: adhesives, cement, concrete, gasoline, fibrous glass, oils, paint, coal tar pitch, poisonous plants, sealers, solvents, sunlight, ultraviolet radiation, wood, and wood preservatives.

3.11 *Cardiovascular effects*

Chemical can affect the cardiovascular system in at least three ways: cardiac sensitization, vasodilatation, and atherosclerosis. Cardiac sensitizers make the heart sensitive to compounds known as sympathomimetic amines, which stimulate the heart to make it beat faster. Vasodilators cause the blood vessels to expand, which results in a decrease in blood pressure and in the amount of blood reaching the organs. Atherosclerosis is a chronic degenerative process in which plaques form on the inside of arterial walls, making the arteries narrower. Many of the chemicals and agents associated with adverse effects on the cardiovascular system have effects on other systems and functions of the body, however, they have been implicated as hazards to the cardiovascular system on the basis of animal and human evidence.

3.11.1 *Examples found in construction*

Materials found on construction sites which have been shown to present potential cardiovascular risks are antimony, carbon disulfide, dichlorodifluoromethane, dichlorotetrafluoroethane, trichlorofluoromethane (used as refrigerant gases and blowing agents in foam systems), nitroglycerine, and pentacholorophenol (a preservative).

3.12 *Liver/kidney effects*

Chemicals which affect the liver and kidney cause their damage by a set of mechanisms which are not well characterized. In most cases, damage to these organs is progressive, as it is related to the intensity and duration of dose. Especially in the case of the liver, small amounts of damage are reversible if exposure ceases, as the liver has substantial regenerative capacity. Many of the agents capable of liver damage cause an abnormal accumulation of fat in the liver cells, which becomes apparent as increased liver size when severe fat accumulation has occurred. At high doses, substances that damage the liver cause cell death leading to tissue necrosis.

3.12.1 *Examples found in construction*

Chemicals found in construction work which have been associated with liver/kidney effects include Aldrin and Chlordane (pesticides), chlorobenzene (a degreaser), and chlorodiphenyls (also known as PCBs) (Table 9).

Table 9. Compounds linked to liver/kidney effects.

Substance	Total exposed FTE (person years)	FTE exposed above permissible limit (person years)
Hexone	24830	200

3.13 *Sensitization*

Allergic responses, also known as sensitization reactions, are the result of an adverse response to a chemical (or other agent) following a previous exposure to that substance, or a structurally similar substance. Sensitization results when the immune system reacts to a substance. The initial exposure to the substance does not produce an immediate response, but that exposure results in the induction of antibody formation; subsequent exposure of the now 'sensitized' individual results in an immune response – that is, an antibody-antigen reaction(the antigen is the allergen in combination with an endogenous protein). This immune reaction may occur immediately following exposure to the allergen, or it may be a delayed response (Lippmann 1997).

In humans the most common target organs are the eyes (the condition is allergic conjunctivitis) and skin (allergic dermatitis). The respiratory system may also become sensitized, resulting in allergic bronchitis and asthma. Once a person is sensitized, that condition frequently persists throughout the lifetime. Symptoms of sensitization may not be observed, however, when exposure to the sensitizing agent, or a structurally similar substance is discontinued. In some cases, however, residual symptoms may persist, as has been reported for workers sensitized to toluene 2,4 diisocyanate (TDI).

3.13.1 *Examples found in construction*

Materials found in construction which are capable of sensitization include plant material (particularly poison ivy, oak and sumac), cobalt metal (dust and fume), platinum, and toluene 2,4 diisocyanate (TDI used in polyurethane foams).

4 HEALTH HAZARDS BY CONSTRUCTION ACTIVITY

The job of developing a comprehensive discussion of the health hazards in construction seems almost as difficult as the task of organizing construction activity itself. The wide range of materials used in construction, the ways in which these materials are manipulated, and the diverse skills which construction workers bring to the worksite produces a complex set of exposure situations. When the fast pace of construction activity is overlaid on this, the result is a complicated, dynamic working environment in which potentially hazardous conditions can appear, change in form, and dissipate before they are even noticed. The resulting complexity in exposures can be approached by classifying construction work and the resulting potential hazards by organizing construction work into the major trades and crafts of the workers who conduct the activities, then link the most significant health hazards to each. The result of this approach is a comprehensive assessment of health hazards associated with construction work, identifying a set of exposures which result from work activities commonly performed by one or more of the trades involved in construction.

4.1 *Abrasive blasting*

Abrasive blasting is a generic term which describes a range of methods used to remove paint, dirt, and other materials from surfaces that require cleaning prior to the application of a protective coating. The exposures associated with abrasive blasting are particulate material generated by the blasting agent and from the surface material. There are four main types of abrasive blasting procedures, based upon the type of blasting agent. Abrasive blasting is done with sand, aluminum oxide, steel shot or slag. When using sand as a blasting agent, workers are exposed to silica (crystalline quartz and respirable dust) as well as high concentrations of nuisance particulates. Abrasive blasting with aluminum oxide can result in exposures to aluminum metal dust in addition to silica and inert dust. Abrasive blasting with steel shot or slag may result in exposures to iron oxide dust. The blasting process also generates exposures to substances contained in the surface material and surface coatings on the material being blasted. The particles produced by blasting may contain lead, iron oxide, chromium, wood, and other materials. In blasting work the crew is likely to be exposed to high concentrations of particulate material. In open air blasting, the process may result in contamination of the general work area as well. In order to control exposures, vacublast units which incorporate local exhaust ventilation at the point of operation can be added. The blasting crew can be protected by sandblasting hoods with supplied air from an outside source. In addition, screens and partitions can be constructed to contain the immediate area of blasting thereby reducing general environmental contamination. The use of dust suppressants, and the prohibition of dry sweeping can also reduce the general level of contamination.

4.2 *Carpentry and glazing*

Workers engaged in carpentry and glazing are exposed to airborne contaminants during the application of caulking compounds and sealant; bonding and gluing; cutting wood; applying wall and flooring materials; and sanding.

Caulking and sealing compounds are typically applied from a manually operated gun.

These materials usually contain a solvent, and may also include heavy metal compounds for mildew resistance. The primary hazard during the application of these material is skin contact which can be prevented by the use of impervious gloves.

Bonding and gluing involves the use of an adhesive which may contain a volatile organic solvent. Typical applications of these materials are not considered likely to produce airborne vapor levels which would require the use of respiratory protection, however, in circumstances where these materials are applied in confined spaces, ventilation may be necessary. Skin protection by impervious gloves is recommended to prevent dermal contact with these organic solvents as well as the adhesive itself.

Sawing of soft and hard wood has the potential to generate significant airborne dust exposures. When sawing is done for extended periods, such as when table saws are set up for repeated, large volume cutting operations, local exhaust ventilation is recommended. The cleanup of large amount of wood dust by dry sweeping can also produce large quantities of airborne dust. Clean up by vacuuming, or the use of dust suppressants can substantially reduce these exposures.

When wall and floor coverings are applied, workers are likely to have direct dermal contact with the coverings, as well as adhesives and grouts. The exposure of greatest significance is likely to be the organic solvents used in the adhesives used to bind these coverings to the floor or wall surface. Local exhaust ventilation may be used to reduce these airborne solvent exposures, and impervious gloves will prevent dermal contact. In cases where exhaust ventilation is not feasible, respiratory protection with organic vapor cartridges is recommended to reduce exposures.

Sanding can be performed by manual or powered methods. In general, hand sanding on wood is unlikely to generate dust exposures of concern. Manual sanding of drywall joints can produce exposures to the calcium sulfate (gypsum) contained in the dried joint compound. Extended periods of drywall sanding can produce excessive airborne dust exposures, especially when the work is done in a confined space. General exhaust ventilation will substantially reduce these concentrations. When sanding is performed using power sanding tools, the potential for dust exposure is substantially greater than in hand sanding. Most powered sanding tools can generally be equipped with built in local exhaust ventilation to reduce dust exposures significantly.

4.3 *Cement, concrete and masonry work*

Workers doing concrete, brick and masonry work are engaged in a range of activities including patching and resurfacing; sealing and coating; grinding, drilling and cutting; handling mortar and cements; preparing surfaces with chemical strippers; installing concrete forms; and mixing mortar, concrete and cement.

Concrete patching materials may contain organic solvents which may present hazardous exposures to vapors, or possible exposure from contact with the skin. Some patching and resurfacing formulations contain epoxy resin systems, which can cause skin irritation and sensitization, and respiratory irritation. These epoxy systems may require the use of ventilation (typically general ventilation where large surface areas are being treated), or the use of impervious gloves and coveralls, as well as respiratory protection certified for organic vapors.

Sealing and coating materials are applied to provide protection against water penetration. Like other coatings, these materials may be applied by brush, roller or spray. In addition to a volatile organic component, these coatings may contain polyurethane systems which are activated by an isocyanate compound. Protection from the irritating and sensitizing properties of these chemicals, as well as the organic vapor component can be achieved through the use of exhaust ventilation, and chemical impervious gloves and coveralls. In cases where exhaust ventilation is not feasible, respiratory protection against organic vapors may be necessary.

Workers may perform grinding, chipping and drilling using saws, hammer drills, chippers, core drillers, and jack hammers. Hazardous exposures to dust, noise and vibration are associated with the use of these tools. While the total dust exposure may be significant, the silica-containing dust presents a particular hazard. High levels of exposure to silica-containing dust have been documented in concrete cutting and surface grinding. Control of exposure to these exposures to silica by wet methods of cutting or local exhaust ventilation is recommended to reduce this hazardous exposure. In cases where these controls are not feasible, respiratory protection is required. Protection against silica-containing dust requires the use of a high efficiency particulate respirator, not the standard single use disposable dust mask typically used for nuisance dust.

Bricklayers, masons, and other workers who mix cement and mortar may be exposed to dusts containing Portland cement and sand. The potential for dust exposure is significant until water is added to the mixture, so the dust exposure is greatest while handling dry ingredients. Workers who conduct trowling and leveling of concrete are very unlikely to have significant inhalation exposures, however, there is a potential hazard associated with dermal contact. A condition known as 'cement dermatitis' is associated with cement exposure. When concrete curing compounds are sprayed onto surfaces, the potential exists for organic solvent exposures. Personal protection in the form of organic vapor respirators may be needed along with impervious gloves to reduce exposure when spraying or handling these curing agents.

Concrete cleaners are used to prepare floors for further surfacing. These cleaners and strippers may contain organic vapors and acids. The corrosive nature of these materials makes the use of impervious gloves, clothing and boots mandatory. Protection from the vapors of these cleaning materials may be possible by exhaust ventilation, however when large surface areas are treated, respiratory protection against organic vapors and acid gas may be necessary.

Workers engaged in the construction of concrete forms may be exposed to oil coating which prevents concrete from adhering to the forms. These oils may contain petroleum distillates, naphtha, or xylene. Exposure may result when these coatings are applied, and when the coated forms are handled. Organic vapor respirators and chemical-impervious gloves are recommended to reduce these exposures.

Batch mixing of concrete and mortar can result in significant dust exposure when handling the dry materials. Specifically, the handling and dumping of dry Portland cement is a significant source of dust exposures. Workers who perform these tasks are potentially exposed to cement dust as well as silica from the dry sand addition to the mixer. Protective equipment in the form of respirators, gloves and coveralls will reduce exposure, however, the potential for excessive exposure to silica in the sand may require the use of high-efficiency respiratory protection rather than the commonly-used disposable dust mask.

4.4 *Demolition*

Demolition is an extremely hazardous construction operation. In addition to the acute hazards resulting from work at heights and being struck by objects, demolition workers are subject to exposure to the full range of materials which were used in the original construction. Notable among these are asbestos and lead, which may not be found in new construction materials, but are widely encountered in demolition of existing structures. The exposures are generally in the form of dusts produced when old structural materials are demolished by explosives, balling, or dismantling.

4.5 *Groundwork and site preparation*

Virtually all construction includes some ground preparation. The nature of the preparation is dictated by the specifics of the construction project, but it is very likely to include the use of some sort of powered equipment. The noise and exhaust gases produced by this equipment can present significant hazards not only to the operators of the equipment but to workers who may be in the vicinity doing unrelated tasks. The construction of a foundation follows the ground works, and the materials used may include cement or concrete. Many types of cement (including asphalt tar, natural, artificial, clay) as well as many types of block and brick are widely used in foundations (and superstructures as well). Dusts can be a major health hazard in cement block, stone setting and bricklaying. Dusts which are generated when mixing concrete and mortar contain silica (including varying amounts of quartz and other forms of free silica), chromium, calcium oxide, and aluminum oxide. Brick and masonry saws and tuckpointers' grinders can generate dusts containing high levels of free silica. Wet sawing does not usually generate significant amounts of dust, however, dry sawing and grinding can create hazardous dust levels.

Cement dusts can cause skin irritation (cement dermatitis), and eye and nose irritation. Mortar and concrete can also cause skin irritation. The cement used in mortar and concrete can produce cement dermatitis from skin contact in its wet (plastic) or dry forms. Some workers may develop allergic sensitivity to certain constituents of cement especially hexavalent chromium. Cleaning solutions used to remove mortar stains from finished wall sections can present hazards. Hydrochloric (muriatic) acid and other cleaners are often used.

Water proofing compounds such as silicone and asphalt-based liquids can present hazards in the handling and application process. Silicone-based compounds can cause skin and eye irritation, and some may evolve hazardous vapors. These materials may require the use of protective equipment and mechanical ventilation especially when they are used on interior walls or in trenches around building foundations. Asphalt-based liquids are applied as both hot and cold waterproofing coatings. These materials are also hazards via inhalation and skin contact, and the hazard they present when applied in trenches or indoors requires engineering controls and personal protective equipment.

4.6 *Interior and exterior finishing*

A wide variety of hazardous materials are used in interior and exterior finishing activities. Flooring materials include a number of types of stone, brick, floorboard, carpeting, linoleum and plastic. Many of these materials are installed with an adhesive which is likely to contain organic solvents. These solvents may cause a range of health effects through inhalation and dermal contact. Some adhesive and mounting systems such as epoxies contain materials which are sensitizers. These materials can cause symptoms to develop in a manner similar to that of poison ivy – no symptoms may appear when a worker is first exposed, but as the worker becomes sensitized through repeated contact, he may suddenly develop allergic-type reaction. Dusts can also be generated in surface preparation for flooring materials, and in final grinding and polishing. These dusts can contain the base materials which includes wood dust, concrete and silica, while the dust produced from the final finishing may also contain the grout matrix, or the hardened epoxy in the case of some terrazzo systems.

Construction of interior walls will involve drywall, plastering and lathing. These activities are conducted using small power tools, and the dusts generated for cutting and fastening the sheets can be a hazard when inhaled. The dusts can also produce dermatitis by drying the skin. Some joint cements can contain silica, which presents an airborne hazard from the dust which is produced when the joints are sanded. Gypsum plaster presents a hazard of skin irritation for the drying effects of the plaster. There can also be a risk of airborne silica dust during the mixing operations. Pneumatic application of plaster coatings can produce high levels of airborne silica in the plaster droplets and dust.

Insulation and acoustical work can present a range of respiratory hazards to the installers. Dusts are released from cement, sand and vermiculite; fibers are generated by mineral wool, cellulose and fiberglass, and vapors are produced by foam insulation systems. These materials can present hazards to many workers beyond the original installers. Workers in many trades which move onto the site later in construction may be exposed to these insulation materials as they install utilities, cables, duct work, and other systems.

Painting and other surface coating tasks can expose workers to a variety of health hazards. Some materials which are no longer in common use such as lead, mercury and cadmium in paints may still be encountered during the preparation of previously painted surfaces. The dusts and fumes of lead, for example produced in the removal of old coating by grinding, scraping, or heat removal can present significant inhalation hazards. The new coatings may contain a volatile carrier (the vehicle) which evaporates from the painted surface. In oil-based systems, these are organic solvents which can be a hazard when inhaled. In water-based latex systems, water is the primary vehicle but there is an organic fraction which is still present in these paints. This organic material may be one of the chemical family known as glycol ethers which serve as emulsifying and suspension agents in the paint.

Provision of utilities includes the installation of systems for electricity, gas, water, telecommunications, heating and ventilation. The health hazards encountered in these trades include those which are primarily associated with the installation, as well as those which the installers contact from work done previously. For example, workers installing electrical utilities may be exposed to metal fumes from the soldering and joining of wiring and electrical connectors, but they may also be exposed to airborne dusts of insulation and fireproofing when they pull cables through channels and openings. Similar conditions are encountered when plumbing and ductwork are installed.

4.7 *Insulation installation and removal*

The application and removal of insulation material involves potential exposures to materials including fiberglass, mineral wool, asbestos, polyurethane foam, urea-formaldehyde foam, and cellulose. Methods of application range from manual laying of batts of material or wrapping pipes, to blown in application of fiberglass or mineral wool. The foam insulation systems are applied using methods similar to the polyurethane roofing systems previously described. Urea-formaldehyde based insulation systems are no longer in use, but this type of foam may still be encountered in insulation removal. In general, the manual methods of insulation installation result in much lower exposures than the blown in methods. These blown in applications can produce high levels of airborne particles making respiratory protection necessary for the installation crews. The application of fireproofing materials is similar in many ways to insulation installation. The fireproofing material is typically a mortar or plaster-like material which is applied by mixing a dry powder with water, then applied to the surface by compressed air spraying. Workers engaged in the mixing activities may be exposed to high levels of airborne particulate material including gypsum, plaster of Paris, and calcium carbonate. While applicators are exposed to the same materials, their exposures are generally lower than those of the mixers, as the mixing involves the use of dry powders which are much more readily aerosolized. Respiratory protection is indicated for most spraying and mixing operations, and in addition, mixing stations can be enclosed and ventilated to reduce worker exposures.

Insulation removal activities are divided into those which involve asbestos and those which do not. Removal of asbestos-containing materials is subject to specific regulations requiring controls such as total enclosure, with exhaust ventilation of the work area, wetting of the insulation prior to removal, and the use of respiratory protection. While these comprehensive exposure controls are not required for non-asbestos insulation removal, exhaust ventilation and the use of wet methods of removal and wet sweeping are effective measures to reduce worker exposures.

4.8 *Landscaping*

Landscaping can involve exposure to a wide range of chemical and biological hazards. Poisonous plants and insects can present hazards to landscaping workers, and the wide range of pesticides used to control insects and weeds can be significant health hazards to the applicators.

4.9 *Painting and other surface coating*

The range of activities under painting and surface coating includes exterior painting of masonry, wood and steel surfaces; grinding, drilling and cutting painted surfaces; application of floor coatings; interior coating of metal substrates; painting of interior walls and ceilings; and cleaning and stripping of old coatings. In addition to this range of activities, there are several methods used for application, including brush, roller and spray finishing.

Exterior masonry may be painted by any of these three methods. Typically, alkyd and epoxy-based coatings are used for these applications. These paints generally include a solvent which can present a hazard to the applicator. The magnitude of the exposure hazard is determined by the method of application (spray painting typically generates more solvent vapor than brush or roller application). The nature of the work area (enclosed, open), the presence of natural or forced ventilation, and the work practices of each painter can also be important determinants of exposure. Because of the nature of the painting activity, painters are usually mobile, and the application of ventilation to reduce solvent vapor exposure may not be feasible. In these cases, painters frequently rely upon respirators, however, proper selection and fitting of respirators is necessary if they are to provide the level of protection they are designed for. Specifically, it is not uncommon to find painters using single-use masks which are intended only for nuisance dusts. These masks do not provide any protection against solvent vapors. Painters should use respirators which are NIOSH – approved and certified for organic vapors and paint spray. These devices contain activated charcoal to remove these contaminants and filters to remove paint particles. It is also important to remember that some of these solvents can pass readily through the skin, so skin protection in the form of gloves and coveralls may be needed. The best source of information on recommended personal protection is the Material Safety Data Sheet (MSDS) provided with the paint material.

Exterior painting of wood is typically done using brush or roller methods. The wide range of coating materials applied to wood includes varnishes. lacquers, stains, primers and paints. As is the case for masonry paints, the exposure risk is largely determined by the specific conditions of application, and the composition of the surface coating. The MSDS should be consulted for information about the nature of the exposure hazards, and the protective measures which should be incorporated in its application.

Exterior painting of steel is usually done by spray or brush. Some coatings used in steel paintings are epoxy or polyurethane-based systems. These materials are favored for their durability, but they can present hazards both to the skin and the respiratory tract. As these materials are frequently applied by spraying, the potential for exposure above allowable exposure limits is substantial. Again, the nature of the painting task frequently makes the use of ventilation unfeasible, so painters are protected with skin and respiratory protective equipment. In some cases, it is necessary to use air-supplied respirators which provide breathing-quality air, rather than relying on cartridge or canister type respirators which remove particles and vapors. The choice of skin and respiratory protection should be made after a careful hazard assessment, which includes consideration of the recommendations on the product MSDS.

Workers engaged in grinding, drilling or cutting painted surfaces are at risk of exposure to dusts produced by the mechanical action of the tool on the surface. The most common metals found in these dusts are metals including lead, chromium, zinc and iron. Of these, lead exposure is the most common hazard, and exposure to lead and the other metals should be controlled by the application of local exhaust ventilation. Many tools are available which include low volume-high velocity takeoff exhaust systems to capture the dusts right at the point of generation. This ventilation control not only reduces worker exposure, it also is an effective housekeeping technique. In cases where this type of exposure ventilation is not feasible, respiratory protection may be required to reduce worker exposure to these hazardous dusts.

Floor coating is usually conducted with a paint system which may be alkyd, epoxy or polyurethane-based. As these coating operations are usually conducted indoors, the risk of solvent exposures due to inadequate exhaust ventilation is high. Depending upon the application, exhaust ventilation may be the best choice to control worker exposures and prevent migration of the vapor to other parts of the building. In cases where exhaust ventilation cannot be used, respiratory protection may be the best method of exposure control. As in the case with other solvent based systems, the possibility of dermal exposure as a significant hazard must be considered, and protective gloves and coveralls may be indicated.

Interior painting of steel substrates may be done by brush, spray or roller. Spray application typically produces exposures of paint mist and vapor which are substantially higher than those resulting from brushing or rolling, and respiratory protection for organic vapors may be required. Workers conducting interior painting of walls and ceilings typically use brush or roller methods. The paints used range from water-based (latex) systems, to oil based (alkyd) and epoxy systems.

Surface cleaning and stripping usually involves organic solvents or concentrated inorganic solutions such as trisodium phosphate. By their nature, these strippers and solvents are corrosive, and they can present hazards from inhalation of vapors, as well as by contact with the skin. Organic solvents typically used include methylene chloride, 1,1,1 trichloroethane, and other halogenated solvent degreasers. Respiratory protection approved for organic vapors, and solvent-impermeable skin protection may be required for the cleaning and stripping processes. The MSDS for each product should be consulted to aid in the selection of appropriate skin and eye protection, as well as respiratory protection.

4.10 *Roadwork and pipelines*

The construction of roadways and pipelines involves asphalt laying, blasting and excavation, pipe wrapping, rock crushing, water well drilling.

Workers engaged in asphalt paving are potentially exposed to asphalt fume which contains a number of known or suspected carcinogens. Typically, paving crews include a worker who operates the paving machine, two workers who level the screed, and several other workers who follow the paver to level, roll, and finish the paving material. Exposures to the members of the work crew vary widely between jobs, and the development of a control technology system was the product of a recent collaboration between the National Asphalt Paving Association and NIOSH.

Workers who perform blasting and excavation of material are exposed to dust, as well as the off-gases generated by the explosion. Significant dust exposures can also occur in the preparation of a site for blasting, particularly in cases where drilling is necessary for placing the explosive charges. In most cases, dust exposures can be effectively reduced by the use of respiratory protective equipment such as single use dust masks.

Pipe wrapping presents a risk of exposure form the liquid asphalt used in the coating material. The amount of asphalt fume and vapor is very strongly determined by the temperature at which the asphalt is applied. Exposure control is most readily achieved by keeping the molten asphalt at the lowest temperature recommend by the asphalt supplier. Personal protection in the form of organic vapor respirators and impervious gloves are also recommended to reduce exposures.

Rock and water well drilling can produce high exposures to mineral dust. The specific nature of the exposure is determined by the composition of the rock being drilled. These exposures can be effectively reduced by the use of water sprays. In cases where this is not feasible, single use dust respirators are recommended. Rock crushing presents a similar source of exposure form mineral dust. Water sprays are also effective methods to reduce exposures. In cases where rock with high silica content is drilled or crushed, further measures may needed to reduce exposure. Control methods such fresh-air supplied equipment cabs, local exhaust ventilation, or high-efficiency respirators may be needed to prevent hazardous silica exposures.

4.11 *Roofing and waterproofing*

The range of activities in roofing and waterproofing includes tear-off of old roofing materials, application of asphalt roofing systems, application of single ply membrane systems, application of polyurethane roofing systems, application of modified bitumen roofing systems, and application of asphalt waterproofing.

Roof tearoff involves the removal of old roofing materials that may include petroleum asphalt, coal tar pitch, fiber board, and asbestos. Old roofing is removed using a variety of power tools, and manual prying, cutting, shoveling and scraping. Tearoff is a very labor-intensive and dirty operation, as it is actually demolition of part of the existing roof structure. The principal hazard is the dust generated in tear off, which may contain materials such as coal tar pitch and asbestos. While these materials are not currently installed in new roofing, workers are still at risk of exposure when they remove these old roofs during tear off. The general guidance for exposure control is to keep the roof thoroughly wetted during tear off by using water sprays and wet sweeping methods. Respiratory and dermal protection are necessary to reduce exposure to airborne dust which may contain coal tar pitch and asbestos.

Built-up asphalt roofing systems involve the use of hot asphalt as the cement and sealer for the roofing system. In these systems, roofing paper is rolled out over the roof in the number of layers specified in the contract (typically 4 or 6 layers). Hot asphalt is pumped from the asphalt melting kettle to the roof, where it is mopped on the seams to cement the roofing felt. The entire roof surface is then mopped with molten asphalt. Exposure to asphalt fumes is likely during the mopping process, and throughout the area where the kettle is operating. Asphalt roofing presents a very difficult exposure control problem. Engineering controls to reduce exposures are generally not feasible for the workers who conduct the mopping activities on the roof. Respiratory protection to reduce inhalation exposures, and skin coverings such as long sleeved shirts and gloves are probably the only reasonable methods to reduce exposures, however, even the use of devices such as respirators is problematic due to the physically demanding nature of the mopping, and the fact that it is often done under very hot conditions. Engineering control systems are being evaluated at this time to reduce emissions and fume exposure to the kettle operators, however, these are not widely used. A series of studies conducted on exposure controls concluded that the most feasible methods are to maintain the temperature of the molten asphalt at the application temperature prescribed by the asphalt manufacturer. In practice, many roofers prefer to operate at higher temperatures because the asphalt is more fluid and easier to handle at higher temperatures, but the fume generation rate and resulting exposures are dramatically higher at higher temperatures. It has also been recommended that properly sized and maintained equipment be used to reduce exposures, and that whenever possible the workers remain upwind of the kettle and the molten asphalt on the roof. Workers should also shower and change out of their work clothing every day to reduce the uptake of asphalt through the skin.

The installation of single-ply membrane roofing involves the use of sprayed, rolled, and brushed-on adhesives. These adhesives typically contain solvents such as acetone, n-heptane, n-hexane, toluene and xylene. These adhesives are applied after the insulation layer has been mounted on the roof deck. The membrane is then laid out, and the adhesive is sprayed onto the back of the membrane and workers roll it out. The seams of the membrane are then brush coated with adhesive, then hand rolled. When these roofs are applied in open air conditions, the likelihood of high airborne solvent exposure is not great, however, many of the solvents in these systems can present hazards as a result of dermal contact, so the overall exposure should be minimized by the use of skin protective gloves.

The application of modified bitumen systems primarily involves exposures to solvents and asphalt. These systems are comprised of layers of rubberized or plasticized asphalt sheets, a glass fiber or plastic support and a protective film. These membrane systems are applied by torch fusing, setting in hot asphalt, or setting cold over a primed surface. The primary exposures of interest in these applications are solvents and asphalt fumes which may be generated in the heating (by torch fusing) or the asphalt fumes which arise when the film is set in hot asphalt.

Polyurethane roofing systems are comprised of two-part systems, one part being the isocyanate containing component, and the other a mixture of curing agent, catalyst, solvent and foaming agent. These systems are sprayed on using an application systems with a mixing gun held several inches from the roof surface. The primary hazards associated with these polyurethane systems are the exposures to isocyanate compounds which are sensitizers. Applicators are also exposed to blowing agents, amines, and a variety of organic solvents in the application and cleanup process.

Workers applying asphalt-based waterproofing compounds may be exposed to asphalt fumes and solvent vapors including benzene, ethyl benzene, petroleum distillates (mineral spirits), toluene and xylene. The risk of exposure during water proofing may be enhanced by work done in poorly ventilated trenches and partially confined spaces below grade around foundations. Protection from inhalation of solvent vapors by respirators and dermal exposure by impervious gloves is indicated in many applications.

Roofing has long been identified as a high hazard work. In addition to the acute hazards presented by working at heights, and burns resulting from contact with hot roofing materials, there is a range of health hazards associated with roofing. Starting with tearoff of old roofing materials, power brooming, scratching, and cutting equipment can generate large amounts of dust. These dusts can contain coal tar pitch, asbestos, and silica. The installation of new composite (built up) roofs may involve hot asphalt which presents a hazard to the kettle operators as well as the workers doing mopping and other tasks on the roof. Asphalt fumes contain a number of compounds which are suspected to be carcinogens, and roofers are exposed to these compounds by inhalation, and skin contact. Newer roofing systems such as plastic membranes avoid the hazard of hot asphalt (but workers may still be exposed to the dusts involved in tearoff). The new membrane systems use a range of organic solvents in the adhesives which weld the plastic sheets together, and these solvents present health hazards when inhaled or when they are absorbed through the skin.

4.12 *Welding and cutting*

Welding is a very common activity in construction. The primary activities include welding and cutting on aluminum, carbon, stainless or galvanized steel; burning, cutting or welding on painted surfaces; and lead caulking.

Aluminum welding is usually conducted using inert gas shielded electric arc methods, while cutting is done by oxy-acetylene torch. The air contaminants generated in welding include ozone, nitrogen dioxide, and metal fumes (aluminum, zinc, magnesium, iron, and copper). Local exhaust ventilation is the method of choice for reducing exposure to these contaminants. Where ventilation is not feasible, air purifying respirators with cartridges for ozone and metal fumes should be used for personal protection.

Carbon steel is usually welded using electric arc shielded with carbon dioxide or nitrogen. Cutting is usually done using oxy-acetylene torches. The primary exposure of concern when welding or cutting carbon steel is iron oxide fume. Gases including ozone, nitrogen dioxide, and carbon monoxide are also significant exposures. As is the case for aluminum, local exhaust ventilation is recommended to control exposures; respiratory protection certified for ozone and metal fumes may be used where exhaust ventilation is not feasible.

Welding and cutting on galvanized steel is generally performed by the same methods as used for carbon steel, but the exposures resulting from work on galvanized steel include high levels of zinc oxide fume. Zinc oxide is associated with the set of symptoms known as 'metal fume fever'. These symptoms include metallic taste, throat dryness and irritation, progressing to chills, fever, and pains in the muscles and joints. Exposure control by local exhaust ventilation should be employed when welding or cutting galvanized steel, with respiratory protection used when ventilation is not feasible.

Stainless steel presents a different set of exposure hazards when it is cut and welded. The welding technology used is usually inert gas shielded (argon or helium) electric arc. Because stainless steel contains significant quantities of corrosion-resistant metals such as nickel and chromium, the fume generated when stainless is cut and welded can be a source of high exposures to these metals. Local exhaust ventilation is the recommended method of control for these metal fume exposures.

Burning, cutting or welding on painted steel surfaces is a significant source of hazardous exposures. The fume produced by these activities can contain components of the base metal (iron, chromium, nickel), and metal and gases which evolve from the vaporized coating material (lead, zinc, chromium, ozone, carbon monoxide). Lead is a particular hazard of concern because it is commonly found in old paint coatings, and it is readily vaporized by the heat of welding or cutting processes. There have been well-documented cases of acute and chronic lead poisoning among workers who were exposed to lead in the demolition, reconstruction and repainting of bridges, for example. The general guidance of providing local exhaust ventilation holds in these cases, however, the presence of lead exposures also invokes the OSHA lead exposure standard (29CFR1926.62) which requires that each employer who has a workplace or operation covered by this standard initially determine if any employee may be exposed to lead at or above the action level. If this determination is positive, the standard includes specific requirements for medical and environmental monitoring and protection of lead-exposed workers.

Soldering and brazing are techniques commonly used for joining metals. These techniques differ from welding in that they do not operate at temperatures high enough to melt the base metal, however they can still result in significant exposures to fumes of the filler metals, rosins and fluxes. Exposures found in soldering and brazing include fumes containing copper, iron, lead tin, zinc, and aluminum, as well as pyrolysis products of the rosin. Local exhaust ventilation and personal protective equipment in the form of air supplied or air purifying respirators are recommended to prevent hazardous inhalation exposures. In addition, some fluxes are corrosive to the skin, and dermal contact should be prevented by the use of impervious gloves.

Lead caulking is a specific activity which is used to seal joints in iron pipes. The lead is maintained in a molten state in this process, resulting in potential exposures to lead fume from the heated pot. Local exhaust ventilation along with enclosure of the area where the pot is located should be used to reduce exposure in lead caulking. Exposure to lead may also invoke the requirements of the OSHA lead exposure standard, as mentioned previously.

5 CONTROLS OF HEALTH HAZARDS

The entire topic of controlling health hazards can be considered in the context of a hierarchical approach. It is essential to recognize that personal protection should be considered the method of last resort in reducing the risks found in the workplace. In the hierarchy of methods which may be used to control workplace hazards, personal protection is not the method of first choice. In fact, it is to be used only when the possible engineering controls which reduce the hazard (by methods such as isolation, enclosure, ventilation, substitution, or other process changes), and administrative controls (such as reducing work time at risk of exposure) have been implemented to the extent feasible. There are cases, however, where personal protection is necessary as a short-term, or a long-term control to reduce occupational disease and injury risks. When such use is necessary, personal protective equipment and devices must be used as part of a comprehensive program which includes full evaluation of the hazards, correct selection and fitting of the equipment, training and education for the people who use the equipment, maintenance and repair to keep the equipment in good working order, and overall management and worker commitment to the success of the protection program.

There are several good examples of successful hazard controls in the construction trades. Probably the best known is the substitution of asbestos insulation with other, less hazardous insulation materials. Another case where substitution of materials is the preferable solution is the replacement of silica-containing abrasive blasting materials. Throughout Europe, the use of silica sand for abrasive blasting is being banned except in the few cases where it is essential. In these exceptional cases, wet blasting methods are being used to reduce the hazard. This practice has not been adopted in the United States, however, where silicosis cases continue as the result of using silica-containing blasting materials. Another example of changing the material and the process to reduce a hazard is the case of adding small amounts of ferrous sulfate to cement, reducing hexavalent chromium to trivalent chromium. The result of this simple change, which has been in effect in several European countries, is to dramatically reduce the incidence of allergic dermatitis among cement workers. In countries where this change has been implemented, the added cost is approximately $1 per ton of cement, while the savings in reduced dermatitis is significant. These examples are only a few of the possible hazard reduction measures which can be taken to reduce workplace disease, at a very reasonable cost.

6 SUMMARY

The nature of the construction worksite makes the identification, evaluation and control of health hazards a difficult task. This paper has proposed a strategy which approaches the problem by describing diseases and conditions which have been associated with exposures to health hazards in construction. With this as background, we can identify the most significant health hazards, and methods of hazard control by breaking the work down into its component tasks and activities, and linking the potential hazards associated with each activity.

This strategy is founded upon the observation made in 1942 that health on construction sites is potentially influenced by 'everything under the sun'. Recognizing that no single approach will catalogue the entire range of health hazards in construction, this approach attempts to address this difficult problem by a hazard assessment strategy centered on construction work activities.

REFERENCES

Lippmann, M., (ed.) 1997. 'The More We Learn, the Things Change, the More Things Stay the Same: Notes on a Pioneer in Occupational Safety and Health'. *Applied Occupational and Environmental Hygienez* 12(10): 648-649.

Susi, P. & Schneider, S. 1995. 'Database Needs for a Task-based Exposure Assessment Model for Construction'. *Applied Occupational and Environmental Hygiene* 10(4): 394-399.

Weeks, J.L. (1998). 'Health and Safety Hazards in the Construction Industry'. In: *Encyclopaedia of Occupational Safety and Health* Volume III, Fourth Edition. Geneva, International Labour Office, pp. 93.2-93.8.

CHAPTER 8

Use of IT to help regulate public health and safety in construction

RON SHARPE
CSIRO Building Construction and Engineering, Australia

ABSTRACT: Information technology including knowledge based expert systems, a universal IT language for construction and software systems for evaluating performance of constructed facilities are helping to improve health and safety.

Knowledge based expert systems assist compliance checking in a thorough and exhaustive manner. While they are not intended to replace experts they provide complementary support to both the expert as well as non-experts including around the clock availability, explanations, automated reporting, archiving plus linkages to other reference material and related software systems such as CAD. An expert system (Windloader) for the Australian wind loading standard is described briefly followed by the development and use of the BCAider system for the Building Code of Australia over the past 7 years.

The latest IT advance is the definition of a universal IT language under the ISO STEP and IAI umbrella to ensure the seamless flow of data and information across all sectors of industry. As well as increasing efficiency and accuracy, STEP will also eliminate errors and re-entry of data that currently occur due to manual processing and incompatibilities in current information systems. The International Alliance for Interoperability is developing the IT standard for the construction sector and feeds into the STEP process. One of the current IAI projects involves defining the IFCs for performance based regulations to enable building code checking is described.

The development of interactive remote communication technologies also enables health and safety concerns to be included during design, construction and operation phases. The first to be discussed is the use of CAD conferencing as an aid between designers and constructors to ensure that construction goes according to plan and also that problems are resolved. The system allows live video communication between the office and the site so that CAD drawings, site video clips and face-to-face communication can proceed simultaneously. A second development is for the remote assessment of sewer pipe using a robot equipped with laser, video and down loading data to computers for real time assessment of sewer pipe conditions. Images are processed using artificial intelligence and pattern recognition techniques. The system can help identify deteriorated sewer sections to enable repairs to be undertaken before collapse and also to reduce the need for hazardous manual inspections.

1 INTRODUCTION

The process of design and construction involves careful consideration of numerous factors that affect the health and safety of both the users of the constructed facility and also the workers and other parties during construction. Over time many regulations and standards have evolved to protect all participants. These regulations and standards need regular updating as standards are improved and as new technologies are introduced.

Information technology can assist in helping to ensure health and safety requirements are met at all stages. In particular expert systems and knowledge based systems can help designers, builders, component suppliers and regulators to check that compliance with structural, fire, noise, air and water quality and other standards are met. Some examples of systems which have been developed for this purpose are described below including discussion of their application.

2 WINDLOADER

Many standards are complex and difficult to interpret quickly and accurately, especially by inexperienced or infrequent users. Errors are often made leading to either costly over-design or inferior and sometimes unsafe structures. This was found to be particularly true in a study by Melchers, 1984, for the Australian Standard for Wind Loads on Structures, AS 1170.2.

AS 1170.2 plays a critical role in ensuring that all structures across the nation resist the maximum wind loads that they are likely to encounter during their life. The nation is divided into four wind zones with the milder category 1 zone in the south and inland, while the northern coastal strips which are subject to tropical cyclones are rated as categories 3 and 4. Much of the code was rewritten after the disastrous 1975 Darwin cyclone in which 80% of the city was destroyed with wind speeds exceeding 265 km per h.

Melchers (1984) found users generated significant errors in applying an earlier version of the AS1170.2. In a study involving 26 engineers designing a factory of dimensions 50 × 25 m, variations in calculated external wind forces on the roof and walls averaged between 10–16% with some gross errors in some cases. But in the case of internal pressures the variations were much greater in cases where the internal pressure sign was assumed +ve instead of -ve leading in some cases to double pressures acting in the wrong direction. A majority also ignored pressure reduction allowances. These results are frightening since it is inevitable that in spite of subsequent manual checking, many errors may still get through.

Individual computer based systems may also contain errors unless rigorously tested. One recent study of spreadsheets found that most contained errors and since designers and others often rely heavily on these, they are also prone to making design errors that could threaten health and safety.

Generally computer software that is widely is likely to be better tested for errors than in-house software. Expert systems that explain the logic used to draw conclusions are also more likely to be error free than the 'black box' type of software that cannot explain how results were derived.

In order to test the value and usefulness of expert systems, Standards Australia and CSIRO collaborated in a two year project to convert AS 1170.2 into a microcomputer-based expert system (Windloader) for estimating wind loads on structures (Sharpe et al., 1990). The expert system not only processes the code logic, but also applies complex mathematical algorithms to derive the wind loads. Users are able to check the mathematical results at each step in order to provide verification if desired plus extensive explanation of how each result is derived. The expert system achieved a modest level of use within Australia with over 150 copies sold since its launch in 1989. User feedback was encouraging when it was first released since it was the first time in Australia that an expert system became commercially available. At the time of its release the predominant operating system was MS DOS. There are no plans to upgrade it to MS Windows because of the expense involved and the difficulty in recovering costs. However Windloader provided many valuable insights into the complexities of processing standards and conversion to commercial grade software. Also the introduction of the IAI and STEP standards will enable software developments such as Windloader to be made independent of operating system changes in future.

The development of Windloader posed many difficulties since AS1170.2 code makes extensive use of complex equations and table interpolations. The code also uses graphics to convey complex concepts, and these were difficult to replicate efficiently using MS DOS. Of course if the project was starting today the graphics and operating system would not be a such a problem

A spin-off from the Windloader development was the discovery of more than 20 errors, ambiguities and omissions in the hard copy that escaped detection by the developers of AS1170.2. As a result the standard was amended for the benefit of the community.

3 BUILDING REGULATIONS

The prime role of building regulations is to ensure the health and safety of building users. The regulations often refer to other standards such as those for structural loading and design, mechanical and electrical services, fire, noise and lighting.

While building regulations have generally been prescriptive in the past there is now a strong move to performance based regulations.

Prescriptive regulations are largely based on past practice and describe acceptable solutions such as the minimum height of a room or the minimum area of a window for a given room size. Often alternative solutions are given for particular situations to increase flexibility and choice on the part of the designer. However in spite of this, prescriptive regulations generally restrict new solutions and innovations such as the introduction of new materials and systems. Such regulations often contain outdated past practices which code committees have not got around to updating.

Performance based regulations are being introduced to enable designers and building product manufacturers to innovate and come up with new alternatives. Such regulations specify how the building or a part should perform rather than describing its dimensions, composition, shape, location, etc. As well as providing good alternatives, the latter can also be included in the list of acceptable (or deemed to satisfy) solutions so that other designers can use them if desired.

Prescriptive solutions are usually easier to administer than performance regulations. If the designer complies with the details of the former then that design is acceptable. In the latter case performance must be proved by other means such as testing, expert opinion or computer modelling in a manner acceptable to the authorities.

In both prescriptive and performance case, the regulations are hierarchically structured into a series of clauses sometimes mounting to many hundreds. Many of the clauses cross-reference other clauses in different parts of the regulations making it difficult for inexperienced users to keep track.

Information technology can greatly assist users to correctly find their way through such regulations by ensuring that all relevant clauses are attended to and that costly mistakes are avoided. Where expert opinion is required, it is possible to also include this in attached notes or suggestions. IT will also be able to assist in the case of performance based design where performance can be assessed with other software packages. The latter may be linked or integrated with the checking software package.

4 BCAider DEVELOPMENT

BCAider has been developed for the Building Code of Australia and since its launch in 1991 over 600 copies have been distributed to building surveyors, architects, engineers, academic institutions and other building professionals. It helps users to quickly and accurately check compliance of building designs with the regulations and it runs on IBM compatible PC computers under MS Windows 3.1, Win95, NT or OS/2. It has received 7 awards including an Australian Design Award for excellence in design and performance, the Australian Software Product of the Year Award, an Australian Institute of Building Excellence Award and an Institution of Engineers Excellence Award. It has also been adapted to the Malaysian building regulations and adaptation to the Indonesian building regulations is currently underway.

The BCAider knowledge based system for the Building Code of Australia (BCA) was developed by CSIRO Division of Building, Construction and Engineering in collaboration with the Australian Building Codes Board (ABCB), Jennings Housing and Butterworths. The BCA regulates approximately $40,000 million per annum of building construction and approval delays have been identified as adding significantly to the time and cost of construction. It has also been shown that a 10% improvement in construction productivity would increase the GDP by 2.5% so that improving the speed and accuracy of applying the regulations using software such as BCAider is expected to produce a significant national benefit.

Prototype systems have been initiated in other countries such as New Zealand (Mugridge et al. 1988), the USA (Dym et al. 1988), Canada (Frye et al. 1992), France (Poyet 1989), the UK (Stone & Wilcox 1987), Finland (Kahkonen et al. 1992) and Norway (Mitusch 1989) but to the authors knowledge, few if any have reached significant commercial implementation. Some hypertext systems have been developed by others and while these are useful for the display of text and cross-references, they do not process code clauses to check applicability and compliance. An exception is the BP-Expert being developed in Singapore as part of the Construction and Real Estate Network (CORENET) system (Bp-Expert, 1998). The latter is an exciting concept making use of Internet and the intention is to have all building plans and designs checked automatically both during design and approval by authorities.

After undertaking a market survey (see below) several thousand users and a range of desirable features were identified. The system was designed to meet these user needs with the anticipation that further needs would emerge after the system was put into operation.

The first version of BCAider was released in April, 1991 and so far 700 copies have been distributed across the nation to building surveyors, architects, engineers, code developers and training plus educational organisations. The BCA is amended every 6-12 months and BCAider updates which include these amendments as well as software improvements are released at the same time.

BCAider features include:

- Processing of all 500 clauses of the BCA and State variations;
- Display of reference material including diagrams (from the Commentary);
- Hypertext display of keywords and other references;
- A keyword index to help locate appropriate clauses;
- Display of relevant clauses for a particular building class;
- On-line help and tutorial system;
- An ability to add multiple sets of notes;
- Generation of compliance and dialogue reports;
- User-definable report formats and printing options;
- Display of related product information such as Australian Building Systems Appraisal Council (ABSAC) reports for technical innovations;
- Networking option for a multi-user groups;
- IBM PC compatible computer operation under MS Windows 3, NT or OS/2.

An Autocad interface to allow building design features to be checked for compliance was included in earlier versions but this is being revised to take advantage of the new IAI interface being included in the latest Autocad release. Also earlier versions contain an interface to the Australian Standards on CD-ROM to enable users to 'hot-link' to relevant standards. This interface is currently being updated to allow a smoother interface to the latest Windows version of the CD-ROM.

The major benefits for users include:

- Easier checking of building designs to ensure compliance with relevant clauses of the BCA,
- On-line help and commentary to assist users and provide background explanations plus examples,
- Presentation of the BCA to users in a state-of-the-art computer based format (including use of simple menus, help facilities via hypertext, error checking and colour graphics),
- An ability for designers to explore a wider range of design options by quickly testing out variations (for example, changes to dimensions, exit locations, fire protection),
- Automatic generation of compliance report files for checking authorities for printing and electronic storage,
- Automatic generation of job files which may be stored and later recovered for modification or perusal, and
- Assistance in the education of building designers and checking officials.

There are also important benefits for the code developers including:

- Validation of the logical consistency of the BCA and proposed amendments including incorporation of State variations, and
- Assistance with future restructuring of the BCA into a performance based code by checking for any contradictions, obscurities and omissions.

5 BCAider USAGE AND FEEDBACK

While the adoption by designers has been slower than expected, feedback from one user illustrates its value. The architect design manager at a major national and international con-

Figure 1. Opening banner for BCAider.

struction company has described its use in his company with great enthusiasm. The company was one of the first to adopt BCAider and employs it as part of its Total Quality Management (TQM) system. In particular the company has found that the software provides:

- An ability to check a design fully and to submit a copy (sometimes 50 pages) of the full dialogue and compliance report with assumptions highlighted; the regulators appreciate this because it makes their job easier to check designs, specify amendments (if any) and give approval;
- The regulators with proof that the designers have fully checked the bca;
- An ability to easily incorporate any conditions imposed by the local regulator into the compliance report for other staff to see and also as a record for future designs;
- The bcaider compliance testing and report serves as an audit process for the tqm requirements of the company; and
- An up-to-date reference copy of the BCA (unlike the text copies where updates and State variations are sometimes misplaced).

In conclusion, the design manager stated that 'using BCAider is not difficult; the whole package is extremely user-friendly, and leads you through the whole process remarkably simply; from a software point of view it is a world leader and quite rightly so'.

The predominant usage of BCAider is for checking (60%) and this is mainly by local government. Most of the 250 local government users originally had little previous experience with computers, but appear to have adapted very well with the user-friendly windows environment used by the software. Some local government users report that BCAider frequently identifies several non-compliance errors not picked up in manual checking. Access to references such as the AS CD-ROM, BCA Commentary and ABSAC are cited as particularly useful plus the ability to automatically generate compliance reports. Most all of the original 120 local government authorities who purchased a BCAider licence in the first year of release have renewed for subsequent years thus showing a firm commitment to the technology. Many have since purchased additional copies and network licences.

Twenty universities and colleges are currently using BCAider for training architects, engineers, building surveyors and other professionals and report that it is a good teaching aid. BCAider includes a tutorial, on-line help and index facilities to help users learn the system as well as teaching the correct interpretation of BCA logic.

6 BCAider OPERATION

After starting BCAider the main menu showing the Parts, Sections and Clauses of the BCA is displayed as shown in Figure 2. A Section, Part and Clause may be selected for processing using the mouse arrow, eg. D, D1, D1.5.

BCAider can be configured for any particular State by selecting the bottom left hand box and where there is a variation the clause is marked with an (*). After each clause is processed, the adjacent Status window lists whether or not the building complies with that clause. Clauses may be processed in any order with Section A usually being processed first to decide the Building Class. In the example above a hospital (Class 9a) is being checked for compliance. Once the building class (or set of classes in a multi-class building) is selected, the user has the option of only displaying clauses relevant to the class or classes. The user may add notes to any clause by clicking in the N column and these may be shared by others if the system is networked.

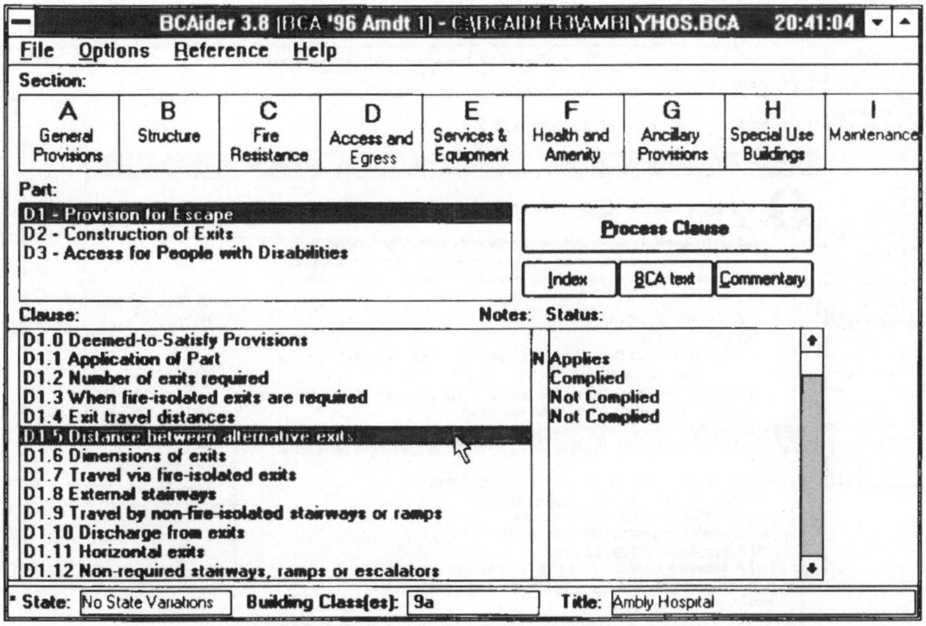

Figure 2. Main menu showing selection of Section D, Part D1 and Clause D1.5.

The 'BCA Index' button in the main menu may be used to quickly find which clause(s) apply to a particular building component. Clause text and expert commentary may also be quickly checked in pop-up windows by selecting adjacent buttons or via the drop-down Reference menu.

Clauses are processed with a series of 'Yes/No' or 'Yes/No/Don't know' questions as in Figure 3. A copy of the dialogue of previous questions is displayed in a background window. Only relevant questions are displayed at each clause and sub-clause level.

Users typically work through the various clauses in the BCA to check if a building complies, and if not, why not. Figure 3 shows a typical screen in which the user is asked for responses regarding building exits in order to compliance with access and egress requirements. Hypertext is used to display underlined terms such as *patient care area* and *exit*. Further expert assistance is available from the Commentary including graphics as shown in a related clause in Figure 4.

Other reference material such as technical appraisals for building products and the text of over 400 related standards on CD-ROM are also available for viewing. The user can also call up other sets of notes generated from previous projects or by other users.

After further questioning BCAider generates a report (which may be saved) stating whether the building complies or not and giving an explanation as in Figure 5. A copy of the dialogue is also provided for the user to check and save if required. Report files may be printed or archived. The compliance checks ensure that users work through the clauses and sub-clauses at a relatively fine level ensuring that no detail is overlooked. While it may be

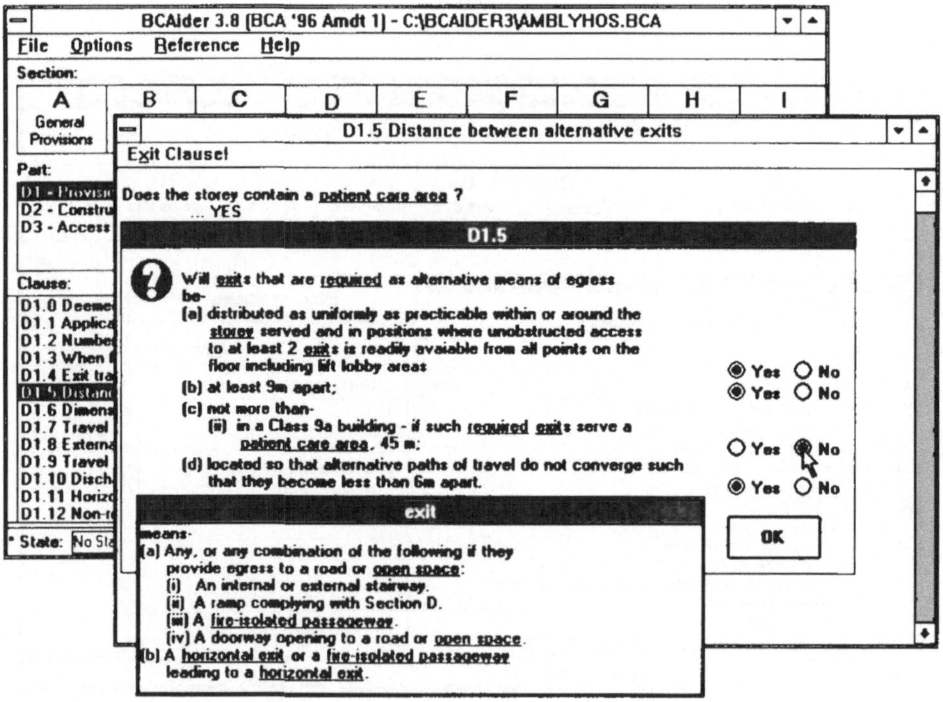

Figure 3. Processing for Clause D1.5 with hypertext used to explain key terms.

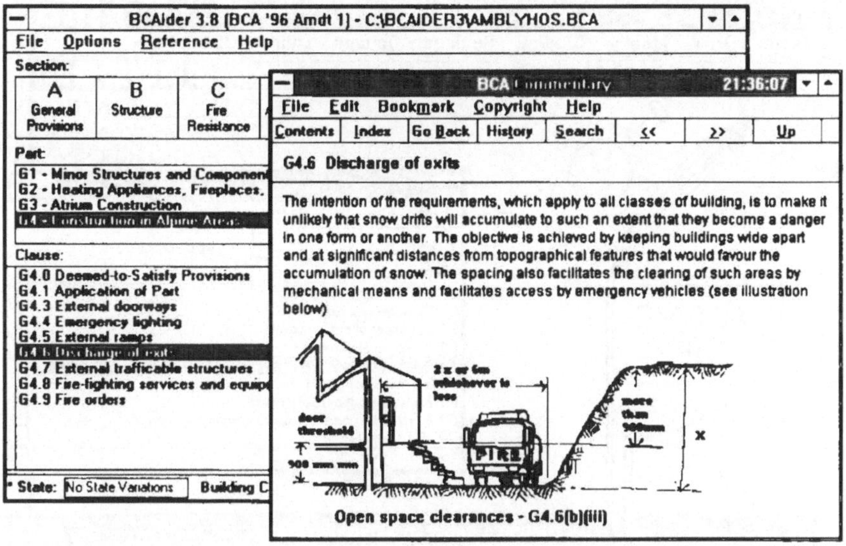

Figure 4. Commentary advice helps users to understand clause requirements.

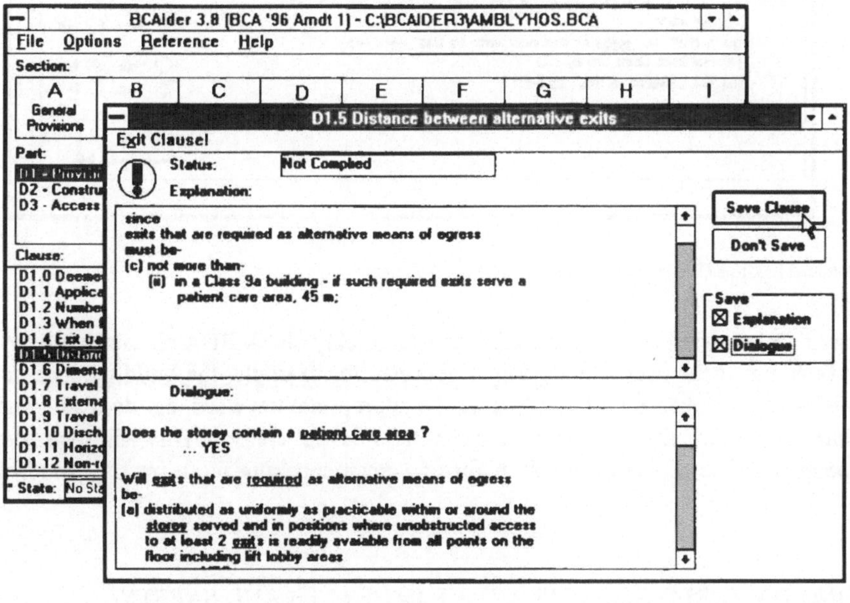

Figure 5. Compliance check for Clause D1.5 with an explanation when compliance fails.

tedious in some cases it can avoid health and safety problems later. For example the author was given a case of a shaft in a major building complex where the fire protection requirements were overlooked at the design stage. The error was not picked up until during construction and added $100,000 to the cost of the project as well as causing an annoying delay.

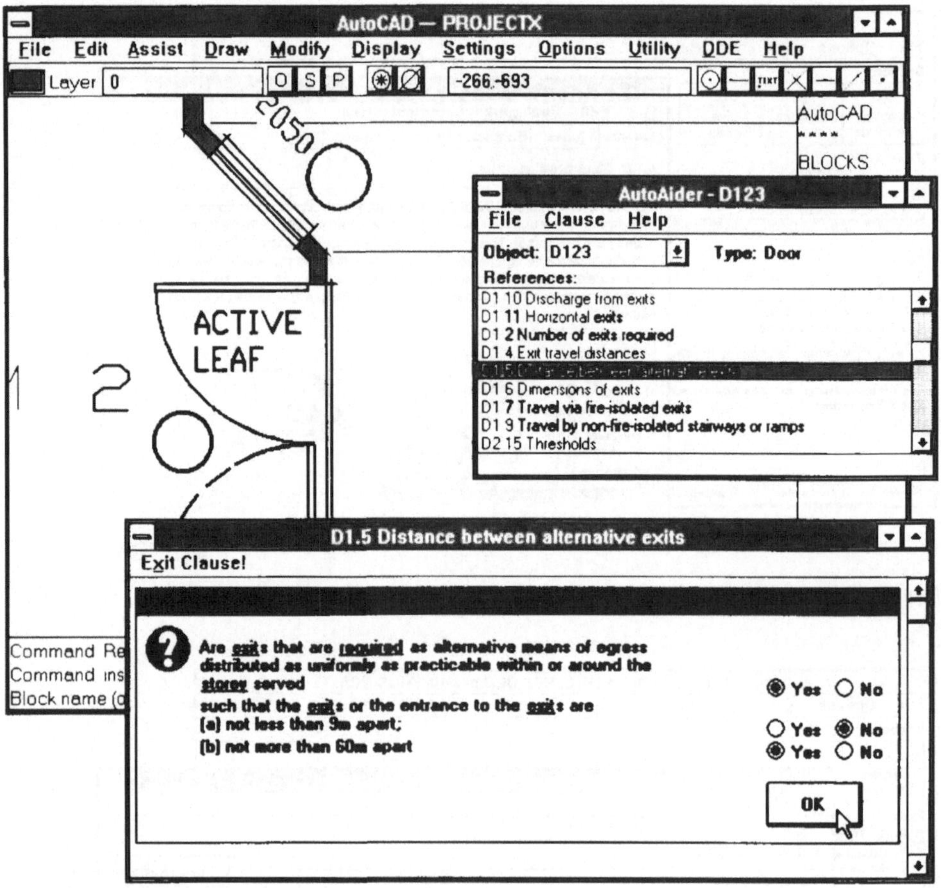

Figure 6. Autocad link to BCAider enables checking of CAD design.

The Autocad (Windows) link enables CAD users to quickly check BCA clauses while designing a building. If the user is designing a doorway for example, the building code requirements can be checked by simply finding the appropriate keyword, eg. doorway, and then an intervening module, AutoAider displays the relevant clauses. This will then take user to the appropriate clause in the BCA for on screen checking in an overlay window (Fig. 6).

7 MOVING TO WORLD'S BEST PRACTICE IN HEALTH AND SAFETY STANDARDS

Several forces are encouraging the internationalisation of health and safety standards and these deserve strong support.

The examples can be clearly seen in the automotive and airline industry where the health and safety of occupants and others have dominated the design process and quickly passed around the globe from advanced countries to the less advanced. Economies of scale of production favour extensive R&D and innovative solutions such as seat belts and air bags in cars. The latter have now become common and well integrated even though they had bumpy introduction phases.

One of the strongest forces favouring internationalisation and best practice is information technology. It can provide the tools which can promote the quality, health and safety in our built environment. It is also driven by cost and productivity needs.

Standards and software are expensive to develop and this provides an incentive for countries to collaborate in their development in groups such as ISO STEP (Standard for The Exchange of Product information) and International Alliance for Interoperability (IAI 1998).

8 IT IN CONSTRUCTION – THE NEED FOR INTEROPERABILITY

Information failures are responsible for many quality (including health and safety) problems and for adding to cost of building and construction projects. Although the AEC (Architecture, Engineering and Construction) industry has made substantial investment in CAD (Computer Aided Design) and IT, effective use of the capital is hampered by lack of compatibility between proprietary graphics systems and databases. Typical problems in AEC include:

- Lack of understanding between project stakeholders (owners, designers, builders, regulators, etc.)
- Lack of co-ordination between disciplines and life cycle stages
- Discrepancies between graphical and textual processes
- Lack of consideration of building processes

The AEC industry has so far failed to assert its requirement for an 'integrated solution' comprising common definitions of objects and functions, supported by software applications written for dedicated use in the sector.

The use of IT in building and construction is seen as a major opportunity to increase productivity. STEP and the International Alliance for Interoperability (IAI) are developing a common universal language for the AEC sector. IAI and STEP are regarded as complementary international R&D initiatives. Formal liaison status of IAI with STEP is being established through ISO. Since September 1995, there have been nine IAI regional chapters established in North America, Germany, United Kingdom, France, Singapore, Nordic Countries, Japan, Australasia and Korea.

The construction industry is traditionally made up of small firms with little ability to effectively integrate their technical and commercial interests. Industry sources say this inefficient exchange of data not only keeps costs unnecessarily high, but is a prime cause of quality problems and cost blow-outs.

The Latham Report in the UK predicted a 30% productivity gain is achievable if interoperable information technology could be coupled with best practice among construction industry partners. If such gains were achieved by Australia's $B 60 pa construction industry, it would mean a 10% boost to the GDP. A similar result is expected for other countries.

Interoperability would allow the seamless flow of consistent and accurate information between all project stakeholders, eliminating data re-establishment between phases and disciplines. This includes design professionals, consultants, contractors, subcontractors, suppliers, facility managers, as well as clients and end users. Singapore through its vision to become an intelligent island, has embarked on an exciting IT project (CORENET) for the construction and real estate sector. Part of its system is to include automated checking of all regulations and standards electronically whenever a project is submitted on Internet (CORENET 1998).

9 INDUSTRY FOUNDATION CLASSES – AN INTEROPERABLE LANGUAGE

Design teams need the capability to develop a design model of the proposed construction on computer from an assembly of 'standard objects.' An object will incorporate physical characteristics (or attributes) such as weight, colour, fire resistance, electrical or thermal properties, as well as representations of how it appears in space in various design and management situations. The attributes may include health and safety characteristics such as strength, flammability, emissions affecting air quality, safe handling and disposal requirements. Process type objects can also be developed for the construction activities such as lifting, materials handling and placement, excavation, welding and concreting. Each of these process objects will have attributes that will include health and safety characteristics.

IAI has agreed on a basic set of project entities that could be used by all groups in a project team. Definitions of these shared entities have been developed as software 'object classes', through a process of collaboration and consensus in which all members of the Alliance participate. These software definitions are now referred to as 'Industry Foundation Classes' providing a common interpretation of real world objects whereby software applications can be developed in a consistent and interoperable manner. The basic set of IFCs are progressively being extended with further attributes as new applications require additional characteristics. For example, the IFC for a wall requires attributes to describe fire resistance properties when a fire modelling capability is introduced.

The IFCs will only provide their full benefits when a large proportion of the construction industry, and the software used by the industry, is based on the IFCs. This will require time and provides one of the reasons for industry joining the IAI – the larger the customer base, the sooner IFCs will be widely used.

Each country is encouraged to join the IAI to ensure that their needs are incorporated in the IFCs, to avoid competitive disadvantage. The IFC R&D methodology currently being implemented can be summarised as follows:

- Through collaboration, co-ordination, and division of labour between centres of excellence world-wide, IFC R&D work is apportioned between countries which participate in IFC development
- Each participating country and its membership belongs to an IAI regional chapter, which draws on expertise and support funding for the part of work it is focused on.
- Each participating country can have access to the outputs from other similar groups around the world.
- Technical integration and conformance testing are jointly developed internationally through the co-ordination of the peak body International IAI.
- IFC development emphasises incremental delivery of interoperable information infrastructure through its IFC object definitions.
- Current IFC work builds on research undertaken by the STEP (ISO 10303 Standards for Exchange of Product Model Data) and other national and international initiatives on

building research and object technologies. Many active IAI Chapter members are prominent figures in the STEP community.

More than 600 organisations around the world are committed to producing and using IFC standard object definitions. Large international companies and research centres such as Archibus, AT&T Network Systems, Autodesk, Bentley, Intergraph, Jaros Baum & Bolles, Lawrence Berkeley Laboratory, Primavera and Timberline are participating in the IAI development for international interoperable solutions. In Australia active partners include Autodesk Australia, CSIRO, NSW Public Works, Lend Lease, Woods Bagot Architects, Construction Information Systems and Peddle Thorp Architects.

For the purposes of developing the IFCs, IAI has divided the AEC sector into various domain areas and teams are formed from interested international member chapters to work on individual domain areas. Proposals for new domain area development must be appropriately resourced and supported by at least 2 international chapters to ensure that the outcome has international application. Where competitive edge is important, it is obvious that countries involved in development of the IFCs will have an advantage in ensuring that the resulting software developed will give them an early adoption advantage over competitors.

As the IFC development process has grown from the work of a small project team to an international collaborative effort it was obvious that full industry coverage by IFCs would take several years. The decision was made to work on overlapping cycles, which would each result in a 'Release' of the IFCs.

The current status of this process is as follows:

Release 1 (R1) – January 97. R1 was demonstrated at the ACS Show in Frankfurt Nov 96. R1 was not intended to be supported by commercial products.

Release 1.5 (R1.5) was officially launched in Singapore in October 97. It is the most important set of specifications to date and will operate over the next 1-1.5 years until it is replaced by R2. R1.5 contains no scope expansion (i.e. it has the same domain classes as 1.0) but has the following improvements:

- Enhanced architecture for upward/backward compatibility as well as for future model extension (independent resource layer, core layer, interoperability layer, domain layer)
- Improved IFC development process documentation (IFC Development Guide)
- Improved model and documentation development tools (OMG IDL specs, meta-model repository, HTML IFC model reference)
- First IFC Conformance Certification Program
- Expanded implementation program
- First commercial applications in 1998

Release 2 (R2) is scheduled for October 98 release. It contains the following major extensions of models:

- Architecture – 65% coverage for typical commercial buildings
- HVAC – 30% coverage of these systems for typical commercial buildings
- Codes and standards – enable some automatic code checking
- Cost estimating – enable early estimation based on a design model
- Facility management – property planning and occupancy planning

R2 implementation is expected to shift away from file-based model exchange toward model sharing via a model server.

Release 3 (R3) started at the Singapore International Technical Management meetings (Oct 97). Key themes of development include:
- Moving beyond data model to standardised object behavior
- Distributed interoperabilty implementation

Domain model extensions include:
- Architecture – 80% coverage for typical commercial buildings
- Power/lighting – 20% coverage of these systems for typical commercial buildings
- Structural – 25% coverage of these systems for typical commercial buildings
- Codes and standards – enable constraint based code checking
- Scheduling – enable early evaluation of schedule alternatives based on design
- Facility management – maintenance, area measurement, systems management
- External libraries – Library access, legal issues and performance attributes of materials for design.

10 AEC CONSTRUCTION STANDARDS – NEED FOR INTEROPERABILITY

The requirement for interoperability also applies to the many AEC standards and regulations that play a critical role in the design and construction process. Such standards need to be better integrated through software into the design process to improve the accuracy, quality and speed of data flows. But at present most AEC standards and regulations are not developed in a form suitable for software integration.

Standards tend to be written by committees with various styles adopted by different parties responsible for different sections. Standards are also amended from time to time often with different styles being adopted once again. Standards often contain errors (numerical and logical), ambiguities, contradictions and omissions. These may escape detection during development since, even with the developers best intentions, it may not be possible to test all possible combinations of conditions exhaustively. Also developers (as experts in their field) may be too familiar with the intention of the standard under development and may overlook the ambiguities seen by the less informed user.

The encoding of standards and regulations into software provides a more rigorous approach to testing the logic as well as exposing other discrepancies since software cannot tolerate ambiguities and contradictions that may be hidden in the source documents. This has been our experience for development of the Windloader software and BCAider as discussed earlier.

If the AEC standards are developed from scratch in the same interoperable language being used in other design and construction software, this would ensure that the standards software would be easier to integrate and also of a better quality. It could also be more readily tested to check for errors and inconsistencies in combination with other design and CAD software. In addition if standards and regulations made use of standardised logic templates, the software task could be made much easier and partly automated.

11 IAI INTEROPERABILITY PROJECT FOR CODES AND STANDARDS

IAI has recognised the importance of Codes and Standards and Release 2 of the IFCs will contain some automatic code checking models. There are two major project areas here:

– Access and Egress for the Disabled in Buildings
– Energy Code Compliance Checking

Access and egress facilities for the disabled is a critical issue now in Western countries and is resulting in major costs to develop solutions to meet the requirements of government legislation to prevent disability discrimination. For example in Australia it is estimated that to comply with new standards it will cost building owners between $4 to $5 billion over the next 15 years adding 5% to 8% to building costs. Obviously this will spur R&D to come up with innovative solutions that reduce the costs. Most of the requirements are related to health and safety provisions for the disables including access ramps, automatic doorways, vertical transport, sanitary facilities, seating, hearing augmentation, alarms, communication and navigational aids. Safety in the case emergency protection such as fire is of particular concern since the disabled needed to be provided with quick egress facilities or safe refuge spaces.

The Access and Egress project within IAI is being led by the Singapore Chapter and is developing the IFCs to enable assessment of escape route provisions and facilities of buildings to serve the needs of wheelchair users and the ambulant disabled. In particular the project will enable checking of:

– Clearance (width, area for wheelchair movement) of space and access
– Surface requirements for floor materials including slip resistance (dry and wet)
– Floor level changes and gradients
– Obstacles in space and access such as doors which inhibit manoeuvrability
– Aid provisions (signage, handrails)
– Escape route occupancy capacity requirements
– Exit requirements (number, location, width)
– Exit approach (smoke free, fire resistance, sprinkler protection)
– Means of escape (refuge areas, stairways and passageways)

The Energy Code Compliance Checking was initiated by the North America chapter with the Australasia chapter recently joining the project. While the project is focused on an energy code, valuable experience will be gained for later application to other code and standards applications.

In the latest development version of this project, IFC model extensions necessary for interoperable performance-based energy code compliance checking applications will be defined. 'Performance-based' here means that compliance is based on overall, whole-building performance rather than on meeting prescriptive design requirements. Performance-based energy codes (or compliance options) are available in virtually all commercial building energy-efficiency codes in the United States and Canada. These codes determine and enforce minimum design standards for building energy efficiency using computer simulations. They offer an alternative to the more prescriptive energy code compliance options addressed by the an earlier project in Release 2. Performance-based codes offer the advantages of greater design flexibility, increased opportunities for design innovation and economic optimization, and compatibility with other design incentive programs; e.g., utility demand-side management programs and the US Environmental Protection Agency's Energy Star commercial building program.

Several performance-based energy code checking applications are now available or under development in the United States and Canada. Historically, performance-based energy code compliance checking has been performed by engineers and architects during the design development phase for large or complex commercial buildings. However, the extensive user input required with performance-based compliance methods have limited their use in most jurisdictions. The capability to acquire pertinent building description data from other design applications will offer enormous benefits in time savings for data entry for those performing this process. This is likely to expand the number and kinds of users employing performance methods and to move use of these compliance tools earlier in the design process – facilitating improved integration of energy efficiency with other design objectives.

12 INTERNATIONALISATION OF BUILDING STANDARDS THROUGH SOFTWARE

As discussed earlier the internationalisation of standards will help improve health and safety standards in construction. Information technology has a major role to play in this process as a result of the:

- Large international installed base of relatively standardised software such as windowing systems for pcs plus standardised general computer languages such as C++.
- High capital development costs for IT systems encourages international focus to ensure that there is a sufficiently large international market to recover costs. Many countries cannot afford the costs of independent development of standalone software.
- Users favour standardised systems as they are generally well developed and tested, likely to last longer, and have greater compatibility with other software.
- Increasing use of graphics and icons cuts across language barriers and make software systems easier to use.
- Networking of computers, especially via Internet, is integrating and effectively forming a single world information system accessible to all. Software developers in any country can readily scan the world and pick out the best ideas.

In the IT area standards largely evolve from the private sector practice and is generally adequately funded from industry sources. But in AEC sector the public sector has been the traditional driver. Many countries developing regulations and standards are seeing less public money being made available for this work and instead are having to raise more funds from the private sector. The latter often comes from the sale of the published documents together with seminars and workshops.

While software is another potential income generator to support standards development, it is usually too expensive for individual countries to take on despite an increasing demand from users. However this provides an opportunity for multilateral collaboration in software developments of harmonised standards thus sharing both the development costs and the revenue generated. For example the IAI program has already recognised standards and codes as a major focus area for development as discussed above.

Thus it is reasonable to project that as IT systems become more powerful we will see standards and codes being developed as software ready for integration in interoperable systems such as those of STEP and IAI.

Ultimately there will only need to be one software copy of a standard or code available on Internet which all can access for automated checking. This is already the aim of the Singapore CORENET system so while it may be a distant dream for many countries, for others it is just around the corner. Imagine the savings in trees and paper, printing, distribution, maintenance and collection of royalties. Internet in a very short time has achieved a common international format, accessible to all and easily understood.

While automated checking of regulations is proceeding, there is also a need for improved computer based processing of standards, especially if advantage can be taken of Internet. Figure 7 shows how a typical clause from a standard might appear on Internet. This displays a clause from the Building Code of Australia in an experimental format whereby the user can see all of the clause, its subclauses and tree structure. It is a dynamic display written in the Java computer language in which the user can select parameters in each box. As each of these are selected or modified the tree structure and compliance advice adjusts dynamically so that the user can readily see which parts are relevant and if compliance is achieved.

The main copy of this clause need only exist in one place on the Internet. When users download it for checking they will always be sure that they have the latest copy. The standards authority will be able to readily collect a royalty for each access. The software system could be designed so that it issues a 'certificate of compliance' for the regulating authority when the project is submitted by the designer. There is no need for paper to exchange hands anywhere.

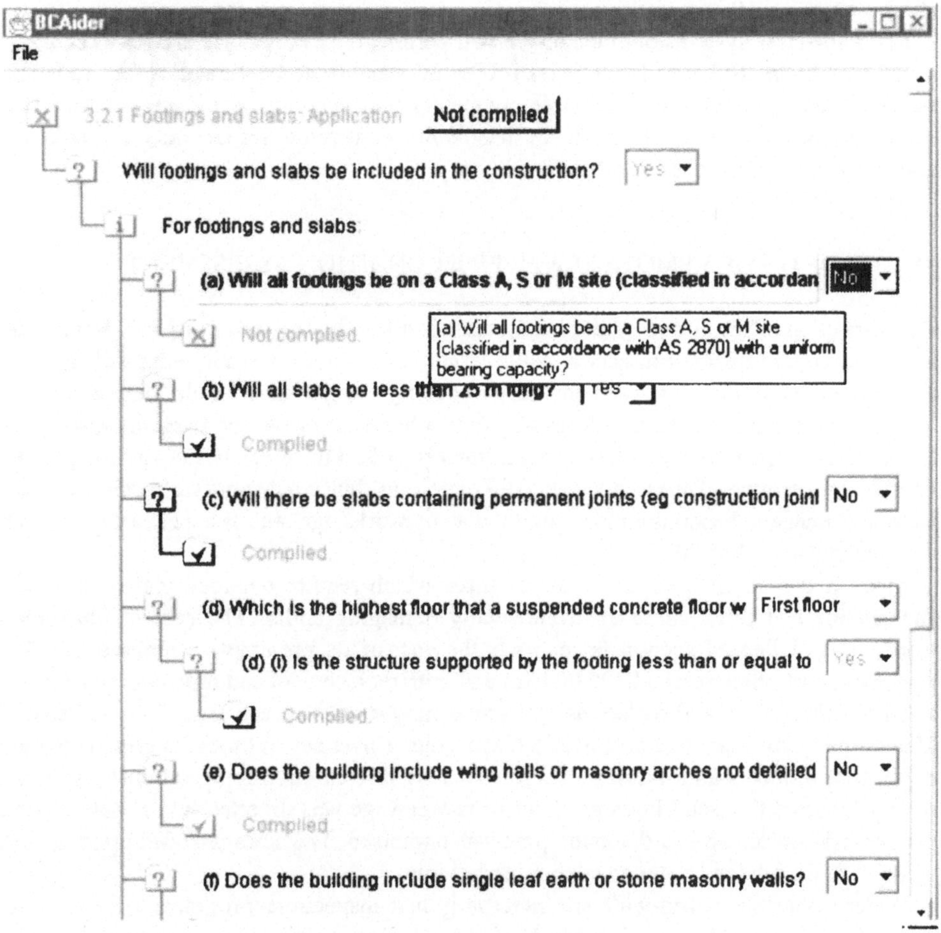

Figure 7. Prototype clause format written for Internet use with dynamic response.

13 REMOTE INTERACTIVE CONSTRUCTION MONITORING

Interactive video and CAD conferencing enables regulations and associated health and safety requirements to be monitored during construction.

As an example, CSIRO and Telecom Australia have demonstrated a CAD conferencing system to prove that CAD/CAM wide area networking is feasible using existing workstation hardware, software and high speed data lines (Newton & Sharpe 1994). The demonstration illustrated how AEC (architect, engineers, construction contractors) professionals can work from different geographic locations with different hardware and CAD systems to collaboratively design, document and construct.

One particular advantage of CAD conferencing is the ability to 'draw or mark-up' on a common screen or 'whiteboard' with each user operating with a different colour pen. Users can either operate on the same drawing or their separate drawings if preferred.

Another advantage is that each users video camera can be used to display other graphical information such as a scene from the site or a video clip. This enables designers, managers and regulators to ensure that actual construction meets the design requirements and also health and safety requirements.

The interactive systems outlined above will permit construction site images to be transferred back to the office and vice versa (office information to be viewed at the site). This process is being simplified as portable computers communicate over digital cellular networks, although there are likely to be bandwidth constraints on interactive image and graphics applications in the short term.

14 USE OF IT FOR REMOTE SEWER PIPE CONDITION ASSESSMENT

The inspection of sewer and other pipe systems is difficult, expensive and very hazardous. Small sewers are difficult to access while larger sewers are often inspected manually by crawling or 'walk-throughs' and this is particularly dangerous with the hazard of flash flooding, poisoning or infection from gas, toxic chemicals, viral and bacterial organisms. Cities in developed countries have approximately 5000 km of sewer per million population. Many sections of these are over 100 years old, but condition is affected by many factors. Damage is becoming more common as networks age, and is accelerating as effluents become more hostile.

Many networks have experienced collapses which require emergency flow diversion and repairs, and often cause environmentally damaging spills. The cost of emergency works for a collapsed section far exceeds the cost of its pre-emptive replacement. Replacement cost often exceed $300,000/km and both replacement and repair costs are rising as other underground and surface assets increasingly restrict access. One sewer collapse in Melbourne a few years ago cost over $20M to repair over several months greatly disrupting traffic, residents and industry. There was a significant environmental cost also when several hundred thousand litres per hour of raw sewage was diverted into a water course over several weeks while a diversion pipe was organised. The sewage ended up in the bay adjacent to the city and polluted swimming beaches.

Water authorities world-wide are increasing their inspections programs to better manage their networks. This is being accelerated by the increasing use of private-sector services by water authorities.

Figure 8. A view from the CAD conferencing system showing how a site details can be compared remotely against design requirements. At the same time discussion of details can be held between site and office staff.

Inspections are generally performed under contract to water authorities by small companies who use specialised video equipment. This is largely supplied by European or US companies with turnovers in the $10m-$50m range. More recently, a number of large Australian and overseas companies have recognised an opportunity for providing more sophisticated and integrated inspection and maintenance services to water authorities.

More than 90% of sewers are 1.2 m diameter or less, and can only be inspected remotely. The most common technique (CCTV) uses a video camera carried through the sewer on an in-pipe vehicle. It provides images to an operator who is required to detect, classify and rate the severity of defects against documented criteria. CCTV requires late-night inspections when most of the sewer is visible above the flow. It is vulnerable to lapses in operator concentration, inexperience, and the inability of the image to clearly reveal important defects. The results are widely agreed to lack reliability and consistency. They provide useful information on gross defects, but lack the reliability to track deterioration so that preventive maintenance can be undertaken with confidence. Preventive maintenance has the potential for major cost savings for water authorities.

Inspection equipment companies have concentrated on enhancing CCTV. This has produced a number of incremental improvements, for example:

- In-pipe vehicles have been made steerable to allow more detailed examinations of chosen areas;

- Light stripe projectors have been provided to assist operator assessment of shape; and
- Scanners for measurement of local features have been trialed.

Other techniques are used for specialised inspections, but they are vulnerable to problems similar to CCTV. A wider range of equipment for these inspections, for example sonar systems for operator-based assessment of flooded sewers, has become available in recent years.

A British sewer inspection company, Amtec, has recently deployed light-line-based (in empty sewers) and sonar-based (in flooded sewers) inspection equipment which allows the measurement of the internal cross-sectional profile of a sewer. Both the sonar and light-line results are in the form of a cross-section at intervals of at least one second, and are recorded on video for subsequent manual interpretation. A German R&D project, Karo, is similarly aiming to enhance existing CCTV technology with new sensors. The Karo technology plans to alert the operator of a possible defect by processing the signals in an intelligent way, however the technology is yet to be proven.

The operational strategy for both current and other techniques under development involves stopping the system for detailed study of defects. This reduces system productivity and is a financial disincentive for accurate inspections under the current charging arrangements which set a fixed price per metre of sewer.

15 PIRAT (PIPE INSPECTION REAL TIME ASSESSMENT TECHNIQUE)

PIRAT has been developed by the CSIRO and Melbourne Water in a joint three-year, $A5m project. In contrast to current qualitative systems (e.g. CCTV), it is a technique for quantitative and automatic assessment of sewer condition (Sharpe et al. 1996).

PIRAT is based on continuous and detailed measurement of the internal geometry of the sewer. The geometry data is analysed automatically to detect, classify and rate defects and the results are presented in a defect report. The results are fully quantitative and repeatable.

Sewer geometry is measured automatically by a custom-engineered instrument system, developed by the CSIRO Division of Manufacturing Technology. The instrument system comprises an in-pipe vehicle and scanner (Fig. 9) which move continuously along the sewer. The scanner repeatedly measures radius as its head rotates around the sewer axis to build up the geometry of a cross-section. Sewer geometry in 3D is established from multiple cross-sections as the vehicle travels along the sewer.

Scanner data is transmitted to a control room in a vehicle on the surface. This houses the processors to extract more than 100,000 radii/m and save them for analysis. A laser scanner is used in low-flow sewers or a sonar scanner is used in flooded sewers where the pipe surface is not visible. The in-pipe vehicle also carries a high quality CCTV facility for navigation and to produce a video record.

The scanning, data processing and communications capabilities are well in advance of those achieved by others and are central to the PIRAT approach and its leads. Their realisation within the constraints of live sewers has been a major R&D achievement.

A defect report and sewer geometry display is shown in Figure 10. This shows the surface geometry data analysed by custom interpretation software which uses AI techniques. The interpretation is based on previously generated geometric signatures of reference defects derived through a pattern recognition learning process. The sewer geometry is analysed to compute signatures of regions which are compared to the learnt signatures, to detect and classify defects and rate their severity. The defect report presents these results

Figure 9. PIRAT in-pipe vehicle and laser scanner.

and individual entries are used to access more detailed information on a defect. Figure 10 shows a typical report segment in which a defect has been highlighted. This has displayed sewer geometry near the defect as a cross-section and an 'unwrapped' map, in which radius is indicated by colour. The interpretation system identifies and analyses the complex and overlapping defects from the geometry data. It is the first known application of these techniques to this task. When coupled with the ability to measure the geometry data, it provides a unique quantitative and automatic system for sewer inspection. The report received by the sewer manager focuses his attention on the few sections which require detailed consideration. The display facilities can be used to confirm the defect report and assess remedial options.

PIRAT has been tested by inspecting 4 km of 600 mm sewer in Melbourne. The testing concentrated on concrete and vitreous clay (VC) sewers and low-flow conditions, so the laser scanner, which has higher performance, was used. The results were compared to independent assessments by an expert CCTV inspector and the discrepancies resolved individually using the CCTV records and geometry data. The system detected all defects found by the expert, as well as a few major ones he missed due to concentration lapses. It

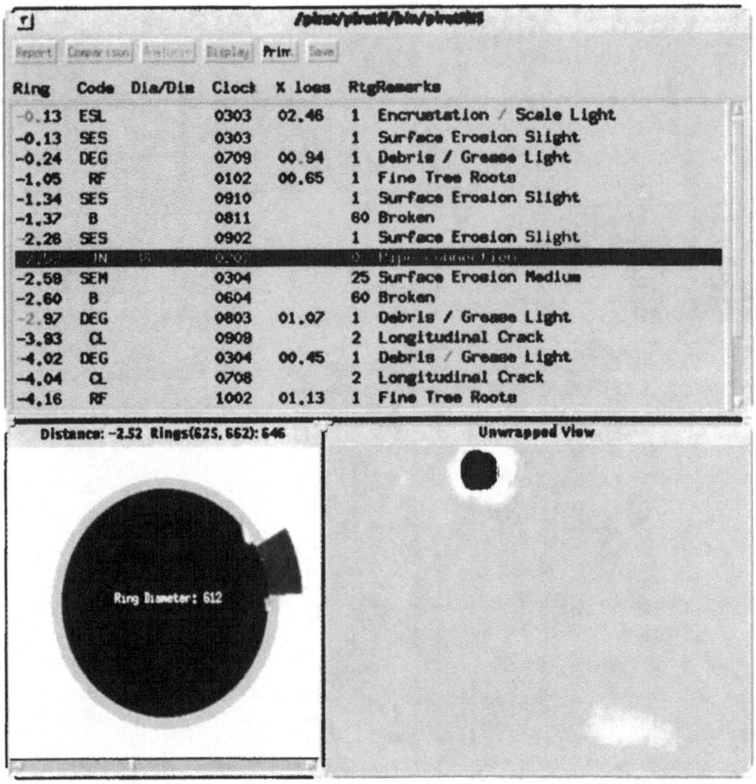

Figure 10. Display of condition assessment reports from PIRAT including pipe cross-section showing a pipe connection.

also indicated several false defects which are readily eliminated manually as they are associated with anomalies in the geometry data.

Figure 11 shows the geometry data for a 4 m section of damaged brick sewer. The 'unwrapped' map contains regions near the top of the sewer and near the middle and right of the map. These are missing bricks. The sharp colour changes between adjacent rows of bricks indicate sharp radius differences and displacement of the brick rows. The region at 40° and the right of the map indicates low radius, i.e. bricks displaced into the sewer. The cross-section confirms this and clearly shows the distortion of the sewer. CCTV inspection would have identified this damaged section, but would not have provided quantitative information on sewer shape. This information has proved valuable to Melbourne Water in evaluating repair options for this section of sewer. All of the above PIRAT images can be viewed in colour on the CSIRO Web page (PIRAT 1998).

16 CONCLUSION

Information Technology in the form of knowledge based systems, remote interactive communication and automated data collection and analysis systems provide exciting op-

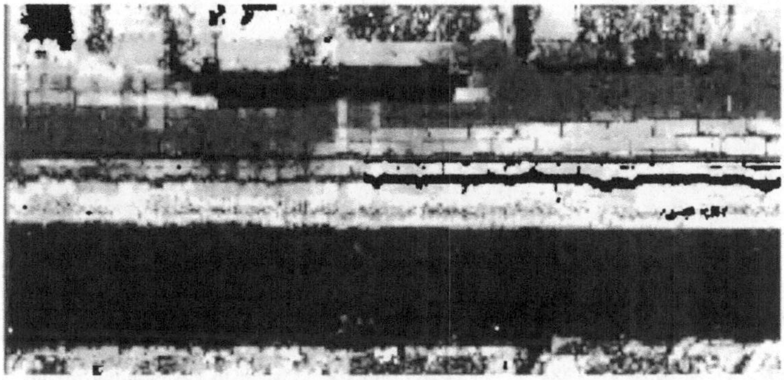

Figure 11. PIRAT image of a damaged brick sewer.

portunities to improve the productivity of all segments of the construction industry. They also opportunities for improving health and safety for both workers and users of constructed facilities.

BCAider has achieved significant acceptance by many building surveyors, architects, engineers, fire authorities and other building professionals in building design, checking and training. It is thus helping to improve the accuracy, ease and speed of applying the BCA and contributing to much needed productivity improvements in the building sector. It ensures that health and safety requirements are met within buildings. A BCAider shell has been developed to enable the system to be applied to other sets of regulations, such as other health and safety regulations. The system has been adapted to the Malaysian building regulations and adaptation to the Indonesian building regulations is currently proceeding. In time it is hoped that it will contribute to improved safety in those countries and avoid many of the disastrous fires and structural failures that occur frequently. In the case of Indonesia, CSIRO is also helping the Indonesian government to revise the Indonesian Building Code.

The IAI developments will enable the smooth and accurate flow on information and data across all sectors of the industry. The quality improvements expected can increase health and safety and all readers of this paper are strongly encouraged to take part in IAI and driving it to help achieve this outcome.

The advances in automation and robotics will enable both users and workers in construction to avoid unsafe environments such as sewer inspection. This is an exciting area of development and heavily dependent on advances in IT.

ACKNOWLEDGMENTS

The author gratefully acknowledges research inputs from Mr Stephen Oakes, Mr John Crawford, Mr Bernard Wilson, Dr Kwok-Keung Yum and Dr Jacek Gibert.

REFERENCES

Bp-Expert. 1998. Web address: http://www.ura.gov.sg/corenet/corenet/whatis/bpexpert.html#intro
CORENET 1998. Web address: http://www.ura.gov.sg/corenet/

Dym, C.L., Henchey, R.P., Delis E.A. & Gonick, S. 1988. A knowledge based system for automated architectural code checking, *Computer Aided Design* 20: 137-145.

Frye, M.J., Olynick, D.M. & Pinkey, R.B. 1992. Development of an expert system for the fire protection requirements of the National Building Code of Canada. *Proceedings of CIB W78 Computers and Building Standards Workshop*, Montreal.

IAI. 1998. International Web address: http://www.interoperability.com/, Australia chapter Web address: http://www.interoperability.org.au/.

Kahkonen, K., Bjork, B.C., Arola, P. & Hult, S. 1992. Building regulations as a part of a digitised building information system, *Proceedings of CIB W78 Computers and Building Standards Workshop*, Montreal.

Melchers. 1984. Human error in structural analysis, Proc. of Seminar on Quality Assurance, Codes, Safety and Risk, Dept. of Civil Engineering, Monash Univ. pp 91-94.

Mitusch, P. 1989. Expert system for the Norwegian building regulations, CIB Building Research and Practice, N4, pp. 223-227.

Mugridge, W.B., Hosking, J.G. & Buis, M. 1988. FIRECODE: An expert system to aid building design, in *Desktop Planning* (eds. Newton, Taylor and Sharpe) Hargreen, Melbourne, pp. 385-394.

Newton, P.W. & Sharpe, R. 1994. Teleconstruction – An Emerging Opportunity, *Proc IEAust National Conf on Construction and Management*, Sydney, pp. 447-459.

Poyet, P. 1989. Integrated access to information systems, Report from Dept. of Computer Science, CSTB, BP 141, Sophia Antiopolis, France.

PIRAT. 1998. Web address: http://www.dbce.csiro.au/ind-serv/brochures/pirat/pirat.htm

Sharpe, R., Marksjö, B.S., Holmes, J.D., Fitchett, P. & Ho, F. 1990. Wind loads on buildings expert system – WINDLOADER, *Journal of Wind Engineering and Industrial Dynamics*, pp. 1269-1277.

Sharpe, R., Gibert, J. & Gipps, P.G. 1996. Impacts of AI and Optimisation on Design, Construction and Maintenance, 15th IABSE Congress, Copenhagen, pp. 943-950

Stone, D. & Wilcox, D.A. 1987. Intelligent systems for the formulation of building regulations, *Proc. 4th International Symposium on Robotics and Artificial Intelligence in Building Construction*, Technion, Haifa, Israel, pp. 740-761.

CHAPTER 9

Automated technology for construction safety

THEO C. HAUPT & JOHN ALEXANDER
College of Architecture, University of Florida, Gainesville, Florida, USA

ABSTRACT: The causes of accidents include compressed project completion schedules, hazardous work situations, and lengthy exposure to hazards. The complete elimination of the hazards, the avoidance of the effects of existing hazards, and the avoidance and reduction of the effects of hazards on workers have ranked as the most significant among steps advocated to reduce the incidents of accidents. Developments in the fields of information technology, automation and robotics hold the promise to change the nature of construction work, and where they have been introduced have already done so. These changes have impacted, and will continue to impact, the safety of workers on construction sites. Automation would, *inter alia,* completely eliminate the exposure of construction workers to the hazards posed by these activities, reduce the amount or duration of exposure the hazards, or reduce the number of workers exposed to the hazards. Automation and the use of information technology could potentially reduce the number and severity of occupational accidents, injuries, illnesses and deaths in the construction industry by improving safety, quality and efficiency through work process re-engineering. Several examples of automated technology and their application in the construction industry will be examined in this chapter.

1 INTRODUCTION

The National Safety Academy has defined an accident in the United States (1995), as an unplanned and injurious or damaging event, which interrupts the normal progress of an activity and is invariably preceded by an unsafe act or unsafe condition. According to this study, about 98% of all accidents were deemed controllable provided there was a will to do just that. It is for this reason that an accident irrespective of its apparent insignificance should be deemed unacceptable – and one too many.

Recent construction industry related statistics for accidents, injuries and fatalities paint a rather dismal, but unfortunately familiar, picture. In the industrialised nations of the world, accidents now cause more deaths than all infectious diseases, and more than any single illness except those related to heart disease and cancer. According to Joyce (1995), the construction industry has always had the highest incident rate in respect of fatal accidents and serious injuries of all industries. The number of fatalities in the industry are only the tip of the iceberg with thousands of major injuries and even more minor ones resulting in lost time.

The construction industry in the United States of America, for example, employs in the region of 5% of the entire industrial workforce. However the construction sector has generally accounted for nearly 20% of all industrial worker deaths (Hinze 1997, Centre to Protect Workers Rights 1995). In Europe the situation is more serious with the construction industry employing 7,5% of the industrial workforce, accounting for 15% of all accidents and injuries and responsible for 30% of all fatalities (Berger 1998, Bishop 1993). Moreover construction has for many years consistently been among those industries with the highest injury and fatality rates.

The causes of accidents include compressed project completion schedules, hazardous work situations, and lengthy exposure to hazards. The complete elimination of the hazards, the avoidance of the effects of existing hazards, and the avoidance and reduction of the effects of hazards on workers have ranked as the most significant among steps advocated to reduce the incidents of accidents. The hazards which may be effectively be dealt with by automation and robotics include working in confined spaces and inaccessible areas, exposure to toxic fumes, chemicals and materials, biological (etiological) hazards, fire and explosions, oxygen deficiencies, electrical hazards, heat and cold exposure, noise, changing climatic conditions, and handling of materials.

Recent developments in the fields of information technology, automation and robotics hold the promise to change the nature of construction work, and where they have been introduced have already done so. These changes certainly impact the safety of workers on construction sites. Many of the most dangerous, unhealthy and labour-intensive jobs have become targets for automation. The consequences of automation would, *inter alia,* be to completely eliminate the exposure of construction workers to the hazards posed by these activities, reduce the amount or duration of exposure the hazards, or reduce the number of workers exposed to the hazards. Automation and the use of information technology could potentially reduce the number and severity of occupational accidents, injuries, illnesses and deaths in the construction industry by improving safety, quality and efficiency through work process re-engineering (Levitt and Samelson 1993). Safety on the construction site influences the critical project parameters of work quality performance, the length of the construction period and the final cost of construction.

However, it is disconcerting to note that the executives of most corporations in the United States cited the five reasons for investing in information technology and automation as labour savings, improved quality, greater product variety, better customer service and faster responses. The benefits for improving worker safety through the use of information and automated technology failed to impact or influence their investment decisions. Is it therefore no wonder that avoidable accidents, injuries and fatalities continue to occur in the workplace? Worker safety is just not the priority that it should be.

2 BIG CANOPY AUTOMATION SYSTEM

The Obayashi Corporation developed 'Big Canopy' as the first automated reinforced concrete construction system in the world. The system integrates leading cutting edge technologies such as a synchronously climbing canopy, prefabricated materials, automated assembly and computerised management of construction materials. Features of the system include:

- Construction can continue regardless of the weather conditions and the prefabrication of building components ensures high quality standards;

- Reduction of the construction time as the skeleton is constructed while at the same time the finishings are completed;
- Reduction of costs since the prefabrication and unification of materials, automated assembly work and computerised management of materials improve productivity and reduce the total construction cost; and
- Reduction of noise levels; and
- Improvement in worker safety.

The system is made up of key elements. These are:

- The synchronously climbing canopy (Fig. 1) covers the entire floor being constructed while protecting workers from inclement weather conditions providing a comfortable and safe working environment. Tower crane posts serving as four columns support the canopy. Climbing equipment of the tower cranes enables the canopy to rise (Fig. 2).

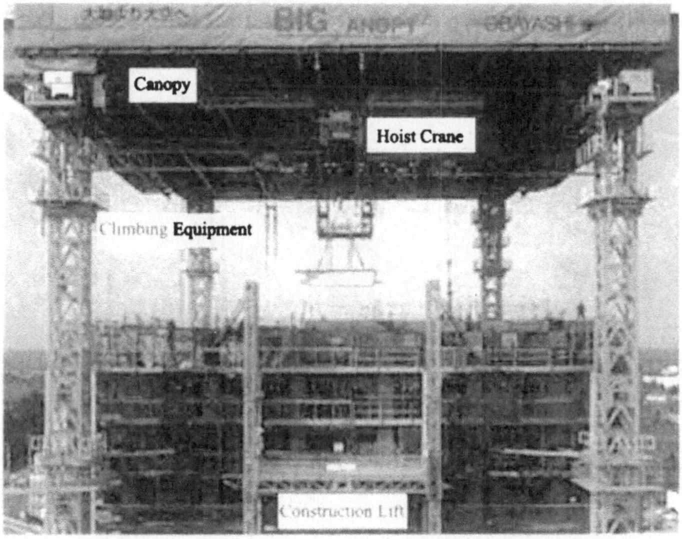

Figure 1. The 'Big Canopy' Automation System (used with permission).

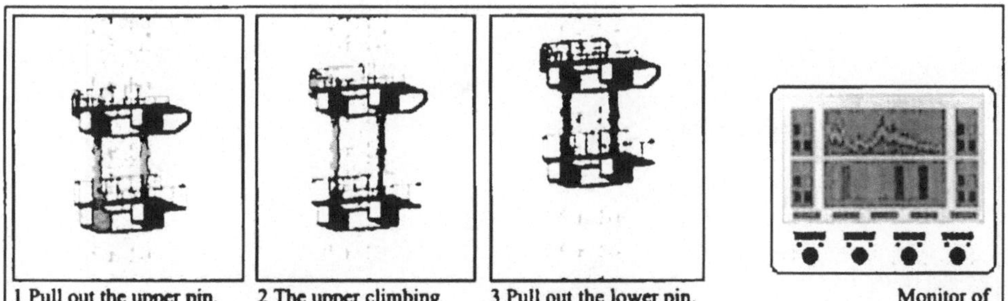

Figure 2. The climbing equipment of the tower cranes (used with permission).

Safety of workers is maintained by synchronised controls.

- Materials are moved by means of a high-speed construction lift and three hoist cranes. Automated hoists move the materials throughout the entire floor being constructed improving work efficiency (Fig. 3). This aspect of the system either eliminates or reduces most of the hazards posed by the handling of construction materials.
- Since most of the building components are prefabricated, only assembly activities are executed on the construction site (Fig. 4). Prefabrication has been suggested as a significant way to reduce accidents on construction sites since in-situ work is considerably reduced and simplified. Work procedures are simple and efficient. The construction process is consequently accelerated reducing the duration of exposure to potential hazards.
- The computer-supported material management system (Fig. 5) reduces the amount of labour and therefore the number of workers exposed to hazards. The system works on the basis of materials bearing bar codes, which provide information about the composition and physical properties and erection location of the materials.

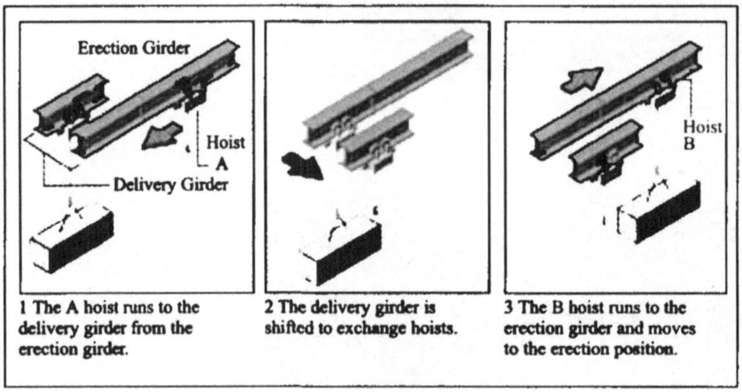

Figure 3. The hoisting system (used with permission).

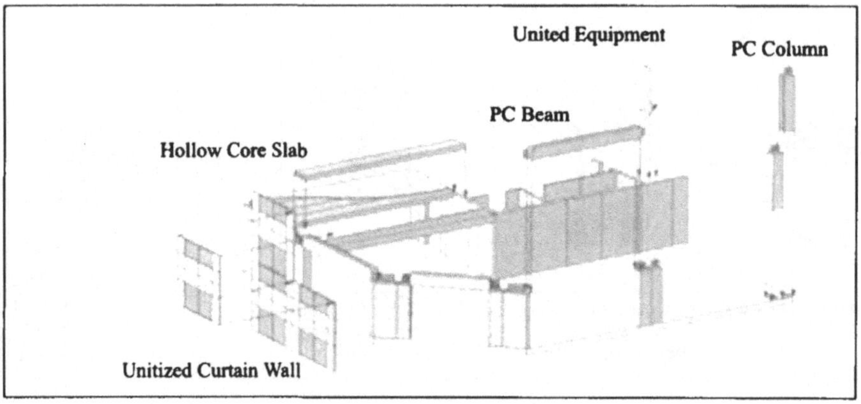

Figure 4. Prefabricated building components (used with permission).

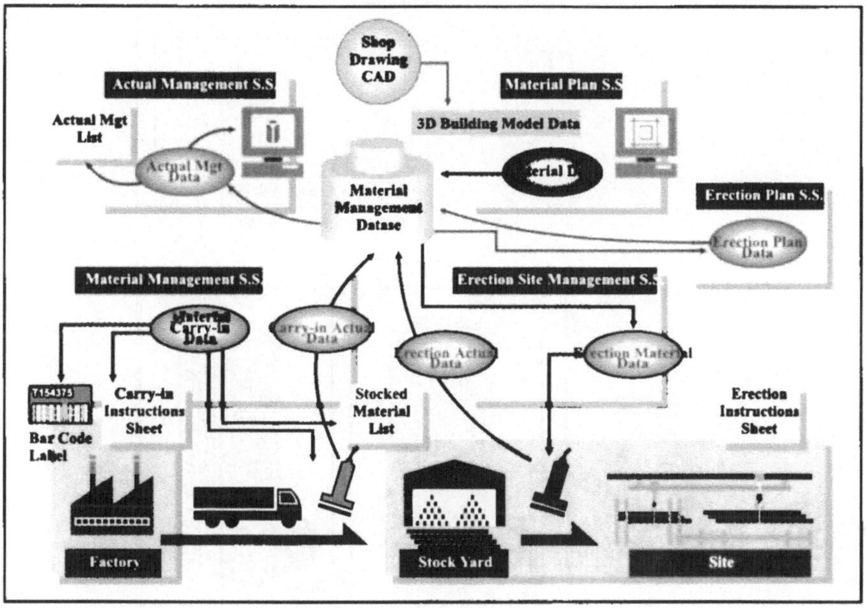

Figure 5. Computerised materials management system (used with permission).

3 SMART COMPUTER INTEGRATED CONSTRUCTION (CIC)

SMART (Shimizu Manufacturing by Advanced Robotics Technology) represents one of the latest attempts at computer integrated construction (CIC) that claims to reduce by 30% the number of man-hours required to complete a multi-storey office building. It operates in a similar fashion to the 'Big Canopy' automation system just described.

System set-up takes about six weeks, after which the top floor of a building and roof are erected on top of four jacking towers, resembling a top hat. The jacking towers are used to push up the 1,323 ton top floor assembly – the main work platform – as well as lifting their own bases from floor to floor in a cycle time of around two and a half hours. The heart of the system is composed of lifting mechanisms and automatic conveying equipment, which is installed on the work platform. This later becomes the roof of the building. Overhead gantry cranes are connected to the underside of the roof structure in a way that resembles a factory production facility. Trolley hoists are used to lift and position components, which are introduced at ground level.

The whole process is computer-controlled, though workers are still involved in overseeing operations at least for the time being. Simplified connections between components facilitate rapid erection times in that self-centring column connections require only fine-tuning with a torque wrench and a laser-guided gauge. A clamp-on welding robot – one of a few task-specific devices – is used afterwards to effect the final mating of the column ends. Floors emerge from under the top hat pre-clad – again from the inside – allowing work in fitting out to begin immediately. Weather is excluded from the job-site by a mesh fabric hung around the work area. Racks of pre-assembled pipework are a further example

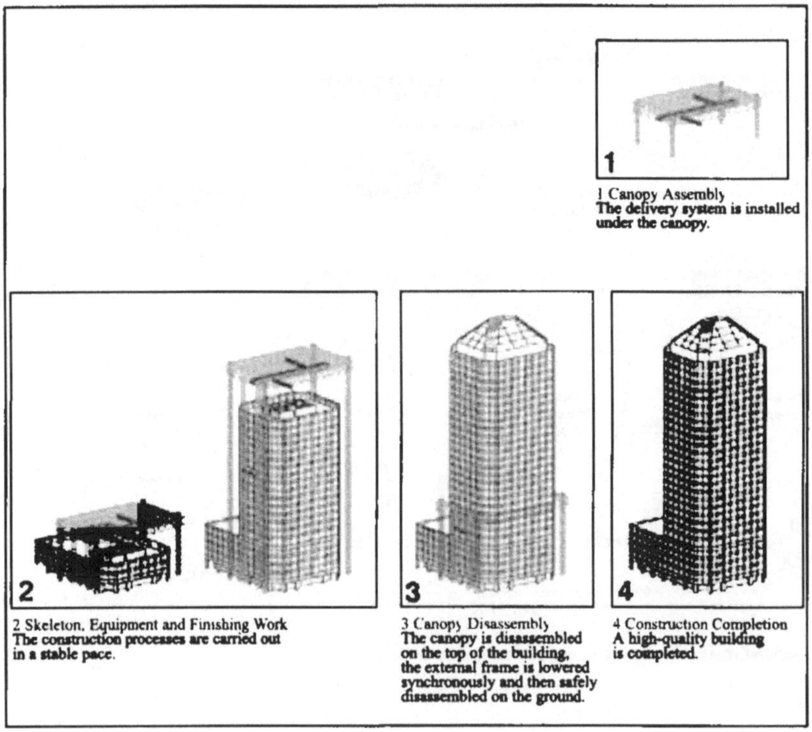

Figure 6. Stages of erection (used with permission).

of an entire approach to rationalising design and production, the aim of which is to drive down the man-hours required for production.

SMART automates a range of production processes including:

- Erection and welding of steel frames;
- Placement of precast concrete floor planks;
- Exterior and interior wall panels; and
- Installation of various prefabricated units.

The SMART system potentially contributes to construction safety effort while enabling the project completion schedule to be compressed by 30%. This contribution is achieved by:

- Eliminating some of the potential hazards on construction sites due to conventional high-rise building construction methods;
- Reducing the amount or duration of worker exposure to hazards; and
- Reducing the number of workers exposed to construction site hazards at any point in time through automation and prefabrication of building components.

4 THE PIRAT SEWER INSPECTION SYSTEM

According to the Commonwealth Science and Industrial Research Organisation (CSIRO), cities in industrialised countries have approximately 5000 kilometres of sewer per 1 million people. In many cases, these sewers are very old with their condition being impacted by age and accelerating as effluents, which flow through them, become more hostile. Additionally, many sewer networks have experienced collapses requiring emergency diversions and repairs. These collapses have caused environmentally hazardous and damaging spills. The cost of these emergency works on collapsed sections of sewer far exceed the replacement cost which exceeds $300 000 per kilometre. Impacting on cost and threat of hazards are the increasing access restrictions presented by other underground and surface assets.

Work in sewers present several potential hazards to the exposed workers, which include:
- Working in confined spaces;
- Exposure to toxic materials;
- Exposure to poisonous fumes;
- Risk of collapse with all the accompanying potential hazards

The PIRAT sewer system has been developed by CSIRO in Australia to provide a system for inspecting and determining the extent of repairs needed in sewers, eliminating the exposure of humans to the hazards referred to.

The artificial intelligence software used recognises, identifies and rates faults and damage, removing the need for highly skilled operators and minimising the impact of human error and concentration lapses that are common during eight-hour inspection shifts. Additionally, the management information system provides clear reports and potentially retains data for accurate comparison with the next inspection, which may take place several years later. These reports provide awareness of the deterioration within a sewer and indicate the possibility of costly faults before they happen.

Other methods provide useful information on gross defects. However, their lack of reliability in tracking deterioration results in preventative maintenance being undertaken with any degree of confidence.

Sewer geometry is measured automatically by a custom-engineered instrument system, developed by the CSIRO Division of Manufacturing Technology. The instrument system comprises an in-pipe vehicle and scanner (Fig. 7), which move continuously along the sewer. The scanner repeatedly measures radius as its head rotates around the sewer axis to build up the geometry of a cross-section. Sewer geometry (Fig. 8) in 3D is established from multiple cross-sections as the vehicle travels along the sewer.

Scanner data is transmitted to a control room on the surface. This houses the processors to extract more than 100,000 radii per metre and save them for analysis. A laser scanner is used in low-flow sewers or a sonar scanner is used in flooded sewers where the pipe surface is not visible. The in-pipe vehicle also carries a high quality CCTV facility for navigation and to produce a video record.

Figure 7. PIRAT sewer system (used with permission).

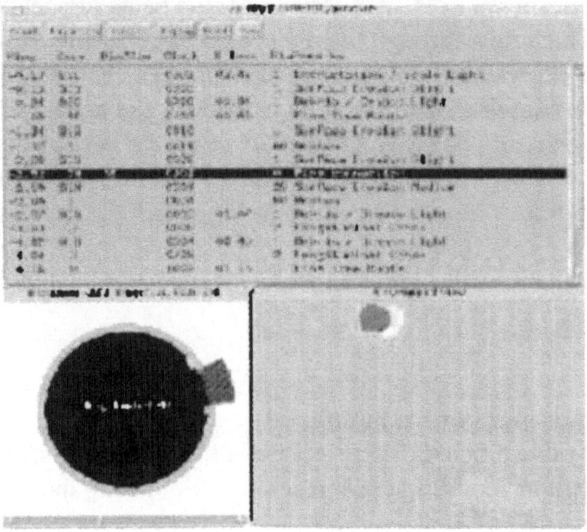

Figure 8. Surface geometry data produced by PIRAT (used with permission).

The surface geometry data is analysed by custom interpretation software, which uses AI techniques. The interpretation system identifies and analyses the complex and overlapping defects from the geometry data.

Inspecting 4 kilometres of sewer with 600-mm diameter in Melbourne has tested PIRAT. Since the testing concentrated on concrete and vitreous clay sewers and low-flow conditions, the laser scanner with higher performance, was used. The system detected all defects found by an independent expert. Further, a few major defects were found which he had missed due to concentration lapses.

5 ROAD AND PAVEMENT MAINTENANCE

New pavement inspection and measurement systems have been developed and an advanced pavement management system has recently been implemented in Sydney. The systems developed by Transfield Pty. Ltd. uses a database comprised of video pictures of every square meter of the pavement being managed. In relation to building and structures maintenance large multi-degree of freedom manipulator arms have great potential for expediting and cheapening structural inspection and condition monitoring work. The exposure of humans to the potential hazards posed by this type of operation is eliminated.

6 AUTOMATED FIELD DATA ACQUISITION SYSTEMS

A neglected and very under-appreciated aspect of construction and management is the topic of field intelligence gathering and analysis.

Many construction-related accidents occur due to the lack of sufficiently detailed information of past accident experiences. This critical historical accident information is therefore not timeously available to the construction manager. Traditionally past accident data are preserved as paper documents and records which are primarily used by head office employees, rather than by staff at the construction sites. This implies that these records are only used for documentary analysis of accidents, and not for accident prevention and to support decisions taken on the construction site. Because of the complexity and the usually voluminous amount of the data, it is extremely difficult for the construction manager to cross-reference the historical accident data with the ongoing project work operations. This difficulty arises due to the lack of appropriate information systems to provide this kind of timely data support to the safety management effort. Until relatively recently, the available analytical and operational tools have been used to conduct safety accident analysis at the documentation level and not in accident prevention efforts.

Geographic Information Systems could be used in safety management in two areas. In the first instance, it could be used in the storage of accident data. The accident simulation process requires data in a multimedia format, which includes the tabular description of the project activity, the geographic description of the activity location and the spatial conditions. Such data can be effectively processed in GIS, with the integration of the multimedia data into one database system. In the second instance, the simulation of accident situations and prevention of accidents is best performed using computerised simulation model processing. The simulation of accidents basically models the configuration of the project, events of the construction operation and predicts the places and activities, which have a high potential for accidents. GIS performs well in the simulation context with the programmer-developed model. A GIS-based accident prevention system could facilitate the analysis of project conditions, modelling of potential accident situations, simulation of potential accidents and demonstrating how the accidents can be prevented from happening.

Such a GIS-based accident prevention system can therefore be used in two ways:

- As a database tool for storing, managing and retrieving past accident data to assist the on-site safety management during the execution stage of a project; and
- As a tool in preventative measures to avoid accidents during the initial planning and design stages of a project in order to avoid accidents during the project execution.
- In general, accident prevention involves multiple objectives and constraints. In Figure 9, what is involved in an accident prevention system is depicted.

Broadly speaking, the project information side deals with the collection, representation, storage, retrieval, and processing of the project activity data. The historical accident database side comprises of acquisition, representation and storage of historical accident data, as well as the use of this data in inference and analysis.

The inference engine, in general, deals with

- The handling of the project information;
- The organisation of historical information; and
- Principles acquired through experience and association.

The engine serves as

- A basis for the analysis of project information,
- A searching tool for relevant historical accident keys, and
- A basis for identification of potential accident locations and situations.

The results of the system are shown in two ways. Firstly, the potential accident area will be displayed on a map. Secondly, the details of the relevant potential accident and appropriate preventative measures will be outlined in the text report.

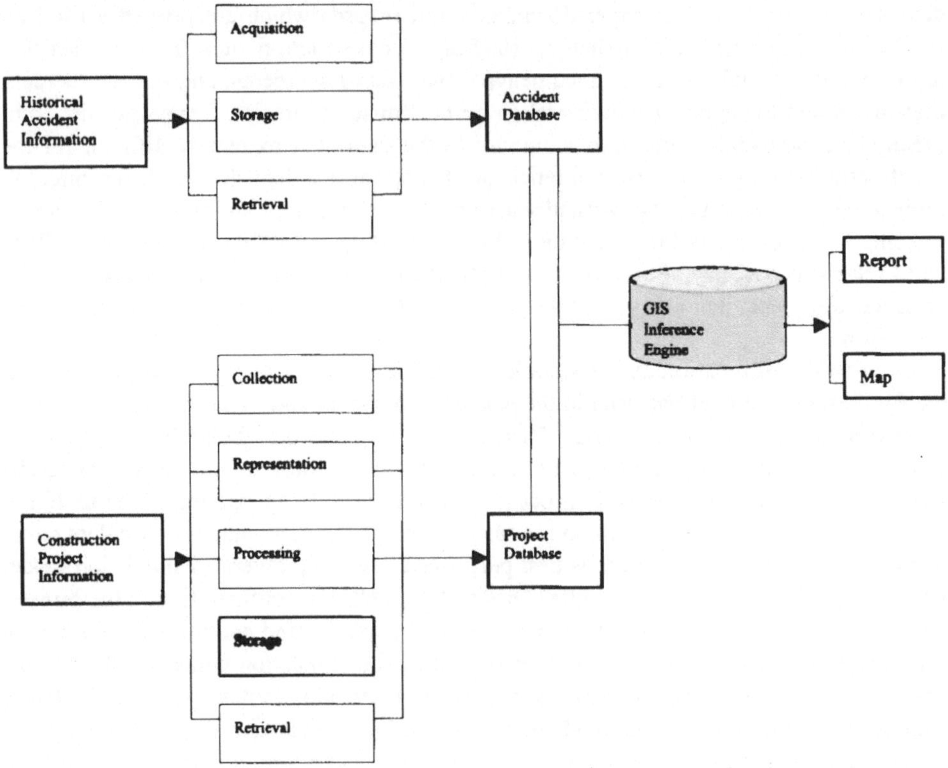

Figure 9. Theoretical framework of GIS-based accident prevention system.

This process of accident prevention can be used to evaluate the safety of the construction project at any stage of the project process. The result of this evaluation will influence the design, planning and construction of the project in order to meet the critical project parameters of cost, time and quality. Using this GIS-based environment, project managers and other project participants can match past accidents which have occurred under conditions and situation which are similar with conditions and situations in the ongoing project and prevent the occurrence of similar accidents.

7 GATOR COMMUNICATOR© MODEL IV PROTOTYPE

Several new technologies, including Global Positioning System (GPS), cellular phones, and digital cameras are being widely used to increase the efficiency of field data collection. However, these technologies, and their associated software and hardware devices, are usually used as independent components of a field data collection system. There have been many attempts by others to connect these devices to handheld computers.

The University of Florida in the USA has conducted extensive research focusing on the integration of GPS and digital camera technologies with handheld computers and Geographical Information System (GIS) software in order to enhance field data collection capabilities. Several other useful component technologies are being integrated into a field data collection device. These include a GPS differential correction receiver, a digital compass, a pitch and roll indicator, digital cameras (for stereo vision) and two-way radio or cellular phone.

Connecting all of these components to existing handheld pen computers or personal digital assistants (PDA's) yields a workable product but one which is difficult and cumbersome to use in the field. Many wires and cables work satisfactorily on a desktop computer but are impractical for a handheld device. This problem confirmed the need for the development of a rugged, fully integrated, easy-to-use, handheld, multi-component field data gathering device.

The Gator Communicator© Model IV Prototype (Fig. 10) is an integrated, multi-component, handheld digital device, which has been designed to increase the effectiveness and efficiency of field data collection and verification. Additionally, it was designed to be an integral component of a real-time, mobile GIS. The device uses stereo digital cameras to measure the distance to points. The capability of the Gator Communicator© to accurately calculate and map the location of inaccessible target objects could be useful in both emergency and non-emergency scenarios.

Typically teams of appraisers collect field data using the Gator Communicator© devices (Fig. 11). The wireless communication capabilities of the device enable them to transmit this data to mobile command centres. Each mobile command centre has the capability to combine the data received from the field team into status, analytical, and sum-

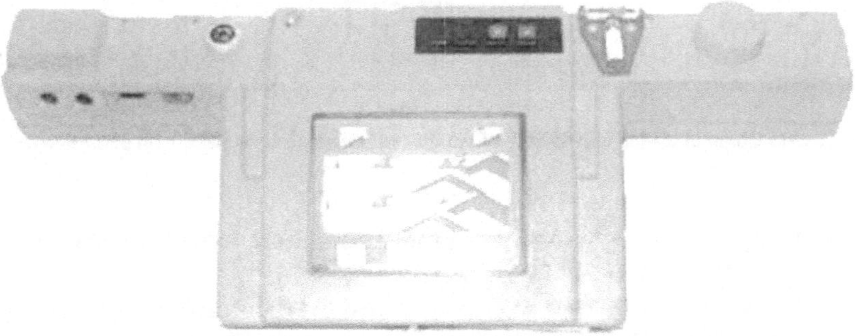

Figure 10. The Gator Communicator© Model IV Prototype (used with permission).

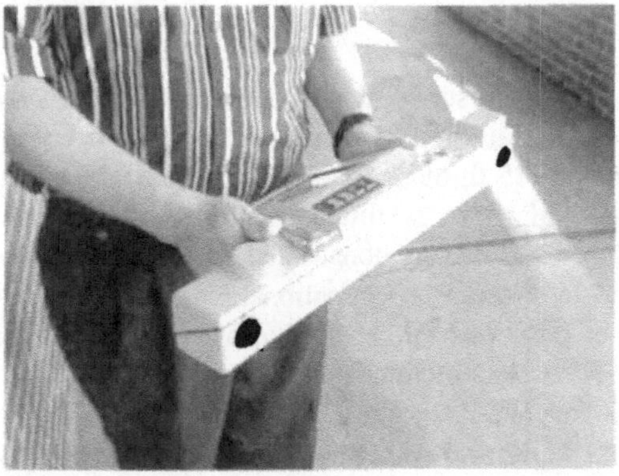

Figure 11. The Gator Communicator© Model IV Prototype in the field (used with permission).

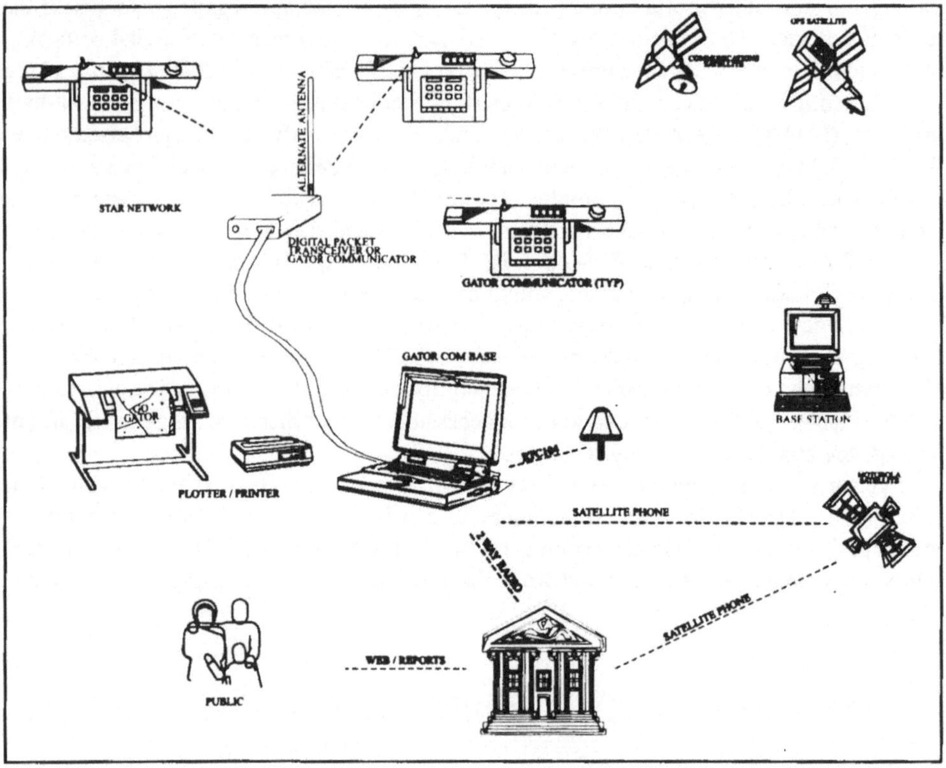

Figure 12. The Wireless Communication Capabilities of the Gator Communicator© (used with permission).

mary reports and maps. These could in turn be printed out for use locally or transmitted to remote sites using satellite or standard phone services.

The World Wide Web could be used to provide access to reports and maps to managers and the general public (Fig. 12). This method would facilitate the real-time data flow from field to the Web and back.

8 AUTONOMOUS MACHINES

Accidents on construction sites involving vehicles resulted in a significant number of the fatalities investigated by the Field Services Division (FSD) of the Victorian Workcover Authority in Australia for the period January 1985 through June 1998. Cranes and earth-moving equipment were the major causes of these fatalities (Victorian WorkCover Authority 1998). Plant and machine operators and drivers constituted 24,7% of fatalities in Australia for the period 1995 through 1996 (National Occupational Health and Safety Commission 1998). In Spain overturning of plant and vehicles rated 5th (3,85%) as the construction activity responsible for fatalities on construction sites.

Driverless and operatorless construction machinery technology has been under development in many countries such as the USA, Japan and Australia. The University of Syd-

Figure 13. Example of large load-haul-dump (LHD) machine (used with permission).

ney and the Centre for Mining Technology and Equipment (CMTE) in Australia have developed autonomous machines which are cost-effective and are being used in more difficult applications such as the operation of load-haul-dump vehicles (in underground operations) and the automated operation of hydraulic excavators and draglines.

Fully autonomous operations of large load-haul-dump (LHD) machines (Fig. 13) for underground applications are presently being developed.

The LHD is able to do autonomous operation using on-board inertial guidance systems, laser ranging, millimetre radar and acoustic sensing systems and has many types of redundant safety systems as well as a transputer based ruggedised on-board computer. Auto-dig, auto dump operations and auto tramming functions are being investigated. The University of Sydney and Komatsu have developed a full sensor equipped hydraulic excavator for autonomous operation testing. The machine is equipped with millimetre radar, inertial navigation and on-board computer systems similar to the LHD machine.

In relation to earthmoving machinery blade control Caterpillar is working with Elphingstone of Tasmania in the development of smart and computer and sensor assisted earthmoving and tunnelling machinery systems. In these systems satellite surveying systems are put on the ends of a grader or dozer blade as position sensors.

The potential contribution of these efforts to reducing the incidence of accidents, injuries and fatalities involving vehicles on construction sites is significant.

9 ROBOTISED PHYSICAL TESTING

The physical testing of hazardous and toxic construction materials can be a very tedious, expensive and dangerous process Many firms have turned to automation technology for improvements in this area.

In Australia, for example, a standard industrial robot (Fig. 14) has been adapted to do a complex series of physical tests of blasted and crushed material in accordance with Australian Standard Specification.

The software programmable machine is able to do most of the normal physical testing activities required – e.g. sampling, sieve analyses, moisture content through oven drying, chemical analyses etc. The results are provided on-line to the head-office thousands of kilometres away. It is evident from this example that industrial robot technology can be simply developed to handle many types of civil and building quality control and inspection processes and laboratory testing operations. The advanced vision laboratories of CSIRO in Mel-

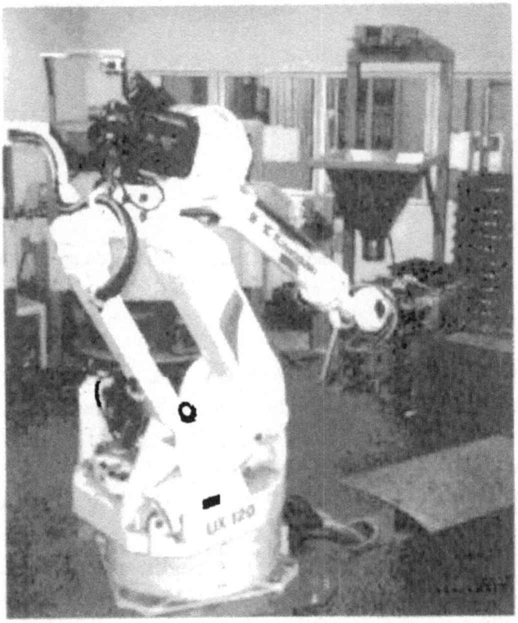

Figure 14. Standard industrial robot used for material testing (used with permission).

bourne have developed advanced vision systems for quality control and real-time process management. At the University of New South Wales advanced close-range photogrammetry systems have been developed for geometry control of manufactured products. These systems yield accuracies of +/–0.4 mm and can handle objects of complex shape.

10 SITE ROBOTICS

Large manipulators (Fig. 15) and the use of high dexterity robotic arms (Fig. 16) are being used to improve the efficiency of on-site construction processes. The University of New South Wales is spearheading the development of this technology in Australia. With the introduction of such technology the materials handling aspects of many construction operations can be much simplified.

Manipulators have been found to be very useful in the execution of reinforcement bar placement and within trench site operations. They also have great potential for safety risk reduction. Time and cost saving of the order of more than 50% in the erection process can be achieved.

11 SIMULATION AND VIRTUAL REALITY PLANNING TOOLS

The use of advanced computer graphics and high speed computing systems have made it possible to do real time, interactive, visual simulation of many forms of construction processes, and to do high fidelity virtual prototyping of products and facilities. This type of technology has the capacity to transform construction operator training and site process planning. The potential for safety training of construction workers without physically exposing them to the actual hazards is enormous.

Since the simulation system is fully general, it can be used to model any form of construction and operations sequence. Such technology is being used now by major contractors in Australia, for example, to explore construction processes and to develop a range of virtual reality planning tools and construction process visualisation technologies for tasks such as process plant erection, crane hook productivity studies and alternative erection methodologies.

12 COMPUTATIONAL MECHANICS

Computational mechanics has emerged as a new discipline involving a co-ordinated blend of insightful modelling of mechanical phenomena with the development of appropriate computational methods. With mechanics, mathematics, and scientific computing as its foundation, the discipline addresses numerical simulation of a wide variety of physical phenomena, with a view toward analysis and optimum design of engineering systems. Incorporated in the engineering system will be aspects of worker safety. Researchers also seek to gain a better understanding of complex physical phenomena which are difficult, if not impossible, to study by alternative approaches. Scientific visualisation tools further aid in this process. The use of ground-penetrating radar (GPR) (Fig. 17) has, for example,

Figure 15. Manipulator (used with permission).

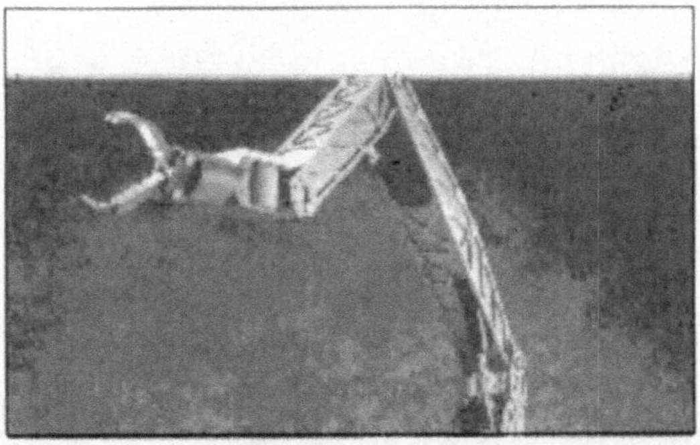

Figure 16. High dexterity robotic arm (used with permission).

been used as a way of inspecting railroad conditions by bouncing electromagnetic waves off ties.

13 AUGMENTED REALITY

Augmented reality refers to the enrichment of the real world with a complementary virtual world. The augmented reality systems, which are being developed at Columbia University, USA, employ a see-through head-worn display that overlays graphics and sound on the naturally occurring sight and hearing of an individual. By tracking users and objects in space, these systems provide users with visual information linked to the physical environment. Unlike most virtual realities, where virtual worlds replace the real world, augmented reality systems enhance the real world by superimposing information onto it. As part of a program aimed at developing a variety of high-performance user interfaces, a testbed augmented reality system was developed that addresses spaceframe construction. Spaceframes (Fig. 18) are typically made from a large number of components of similar size and shape. These components are typically cylindrical struts and spherical nodes. While the exterior dimensions of all the members may be identical, the forces they carry,

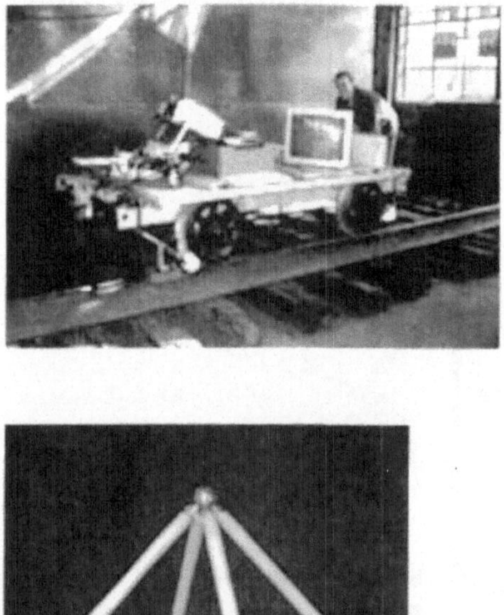

Figure 17. Ground-penetrating Radar (used with permission).

Figure 18. Spaceframe (used with permission).

and therefore their inner diameters, vary with their position in the structure. Consequently it is relatively easy to assemble pieces in the wrong position which could lead to structural failure. The augmented reality construction system is designed to guide workers through the assembly of a spaceframe structure, to ensure that each member is properly placed and fastened.

The system's head-worn display (Fig. 19) is a *Virtual I/O* see-through stereoscopic colour display with integral headphones and orientation tracker.

For each step of construction, the augmented reality system:

- Directs workers to a pile of parts and informs them which part to pick up. Displaying textual instructions and playing a sound file containing verbal instructions does this.
- Confirms that the correct piece has been selected. Selection of the appropriate piece is achieved by scanning a bar-code on the component.
- Directs them to install the component. A 3D virtual image of the component indicates where to install the component, and verbal instructions played from a sound file explain how to install it.
- Verifies that the component is installed by requesting them to scan the component with the tracked bar-code scanner. This checks both the identity and position of the part.

It is evident how the use of augmented reality could be used, for example, to ensure the correct construction of scaffolding systems reducing the likelihood of collapse. The technology has tremendous potential in the area of skills and crafts training.

14 WEARCAM

The 'Wearcam' is an Internet-connected computer which has one or more cameras, microphones, high speed video processing and communications link, as well as its own Internet address. The device (Fig. 20) can be worn naturally while the wearer is performing other duties.

There is a virtual image provided by a miniature cathode-ray tube within a pair of eyeglasses. The camera sees exactly what the worker sees. This device is still in its developmental stages but has the potential to perform the same functions as the Gator Communicator©.

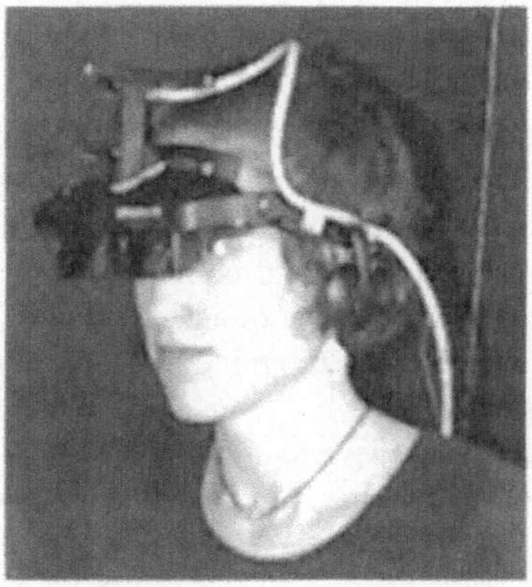

Figure 19. Head-worn display (used with permission).

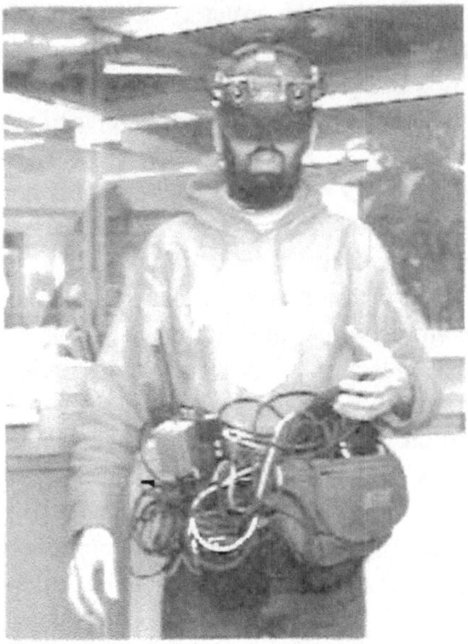

Figure 20. Wearcam (used with permission).

15 LOLA VISUAL OBJECT DETECTION AND RETRIEVAL

This project by North Carolina State University, USA is aimed at the integration of visual object recognition, orientation and retrieval components required for achieving autonomy in mobile systems. When an object is recognised it is possible for it to be picked up and delivered to a predetermined location. LOLA is equipped with sonar sensors and a tactile bumper system for navigation. The sonars determine the range of objects in every direction. Contact on the bumper rails stops the robot to prevent damage to both the robot and surrounding objects. This application could potentially be useful for the retrieval and delivery of construction materials, for example, to areas which are inaccessible, dangerous or polluted eliminating the hazard to workers.

16 THE JENNER PROJECT

Jenner (Fig. 21) is a biped walking robot controlled with Echelon LonWorks field bus technology. It will be able to walk forward and turn, accepting simple commands from a user. It will be designed to be a remotely controlled, versatile walking platform capable of carrying different loads across varied terrain. The walking speed and direction can be controlled. Technology will enable Jenner to remain erect in the event of a power failure. This safety feature will prevent the material being handled from dropping and possibly causing injuries or fatalities to workers in close proximity.

Once Jenner goes into production, it has the potential to retrieve and deliver construction materials and components to inaccessible, hazardous or polluted areas which are dangerous for workers to operate in.

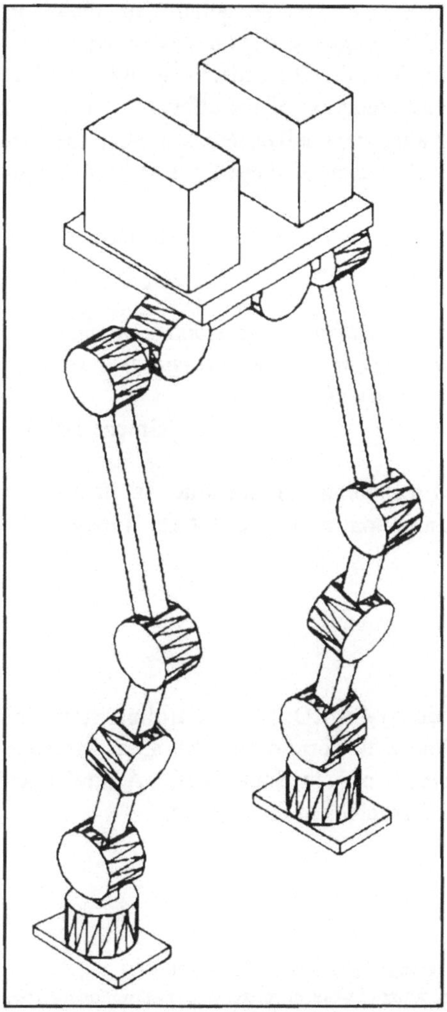

Figure 21. Jenner biped walking robot (used with permission).

17 CONCLUSION

Robotics and automation are developing and changing at an ever-increasingly rapid pace. While the selected automated and robotic technologies described in this chapter only constitute a small sampling of what is available, each has the potential to contribute to reducing accidents, injuries and fatalities of construction workers on sites around the world.

A word of caution!

Before delving into the implementation of automation and robotic technology at a detailed level, each company should first perform a high level of strategic analysis to identify and select an action strategy which best suits the unique preferences and goals of the company. Implementation must be planned very carefully. It may be practical to phase in information, automated and robotic technology through selective automation. This 'phased in' approach must be accompanied by appropriate and thorough training on the new systems.

As has been stated earlier, the complete elimination of the hazards, the avoidance of the effects of existing hazards, and the avoidance and reduction of the effects of hazards on workers have ranked as the most significant among steps advocated to reduce the incidence of accidents. Automation and robotics enable each of these three steps to be addressed effectively.

According to Peter Browne the information technology director for the licensing subsidiary of Microsoft in Reno, Nevada,

> *'Using information technology and its tools, whether through a personal computer or some other type of electronic equipment, is equivalent to driving a car. It is something that many people need to be able to do to cope with the future.'*
>
> (Greene 1998)

The implementation of robotics and automated technology on construction sites to improve the safety of workers may no longer be an option in the near future. It may in fact become a necessity!

ACKNOWLEDGEMENTS

The authors acknowledge the permission granted by CSIRO; North Carolina State University, USA; Columbia University, USA; Obayashi Corporation; Shimizu Manufacturing; Steve Mann; University of Florida; University of New South Wales, Australia; and University of Sydney, Australia to reproduce their graphics in this chapter.

REFERENCES

Allair, William P. 1991. 'Barcoding: standards facilitate integration,' *Automation,* 38(10): 46-47.

Alsene. Eric 1994. 'Computerised integration and the organisation of work in enterprises,' *International Labour Review,* 133 (5-6): 657-676.

Berger, Joachim 1998. 'The Health and Safety Protection Plan and the File containing Features of the Building according to EEC Directive (92/57)' Public Lecture, ME Rinker Lecture Series on Safety and Health in Construction, University of Florida, USA, 20 October 1998.

Britannica Online 1998. 'Safety' *Britannica Online* http://www.eb.com/cgi-bin/g?DocF=micro/516/60.html

Center to Protect Workers' Rights 1995. Program, 2nd National Conference on Ergonomics, Safety and Health in Construction. Washington, DC.

Davis, Joseph R. 1995. 'Automation and other strategies for compliance with OSHA ergonomics,' *Industrial Engineering* 27(2): 48-51.

Greene, Marvin V. 1998. 'Information technology: your pathway to innovative jobs and careers.' *The Black Collegian* 29(1): 112-115.

Hanna, Nagy 1991. 'Informatics and the developing world,' *Finance and Development* 28(4): 45-47.

Hinze, Jimmie W. 1997. *Construction Safety*: Prentice-Hall, Inc: New Jersey.

Horst, Robert L. 1992. 'Investing in automation: total quality management unlocks the dollars,' *Penton's Controls and Systems* 39(1): 46-48.

Joyce, Raymond 1995. The Construction (Design and Management) Regulations 1994 Explained: Thomas Telford Publications: London.

Levitt, Raymond E. & Samelson, Nancy M. 1993. *Construction Safety Management*, New York, John Wiley and Sons, Inc.

National Occupational Health and Safety Commission 1998. Compendium of Workers' Compensation Statistics, Australia, 1995-1996, Canberra.

Rickard, James G. 1994. 'Successful automation begins with basic engineering,' *Industrial Engineering* 26(2): 28-29.

Rifkin, Jeremy 1996. 'Civil Society in the Information Age: workerless factories and virtual companies,' *The Nation* 262(8): 11-15.

Schrage, Michael 1999. 'Finally, an honest assessment (problems facing the technology community),' *Computerworld*, February 8, p. 1.

Singh, Indramani L. 1993. 'Individual differences in monitoring failures of automation,' *The Journal of General Psychology* 120(3): 357-373.

Victorian WorkCover Authority 1998. Statistical Report 1997-1998, Melbourne.

CHAPTER 10

Hazard communications using interactive computer based training techniques

PIERCE JONES
Agricultural and Biological Engineering Department, University of Florida, Gainesville

ABSTRACT: From their inception, computers were touted for their capacity to manipulate, store and retrieve information (data). There is growing recognition that computers can be used very effectively for direct, interactive training. This paper describes a project with two goals: first, to create an electronic version of a hazardous chemicals database that improves access to information and second, to increase the relevance of the database by linking it to frequently used training modules. *HazCom Interactive* runs on personal computers and delivers information on hazardous chemicals from the International Chemical Safety Card (ICSC) database developed by the International Programme on Chemical Safety (IPCS). The ICSC uses standard phrases in multiple languages to describe the characteristics of ~1,200 hazardous chemicals. Pictographs that correspond to many of the standard phrases were created and are displayed to reinforce text. Audio files can be activated to 'read' the textual descriptions of the chemicals. *HazCom Interactive* provides specific, high quality information and is linked to relevant training modules on ventilation, respirators and PPE. The training modules are designed to be highly interactive and generally operate like games. In the project's final phase the software will be evaluated for its efficacy in actual training situations.

1 INTRODUCTION: THE HAZARD COMMUNICATIONS PROBLEM

In the modern world, it is not unusual for people to have a paradoxical attitude toward information. On the one hand, daily life's complex interactions seem to require ever-increasing access to specific, accurate information. On the other hand, there seems to be an expanding overabundance of information. As John Naisbett said in his 1982 book, *Megatrends*, 'we are currently drowning in information, but we are starved for knowledge.' The dilemma is how to efficiently sift through all of the unwanted information to access precisely what is needed, when it's needed. Naisbett's observation certainly rings true in the area of occupational safety and health as illustrated in the following example.

In 1983 (the year following publication of *Megatrends*) the US Occupational Safety and Health Administration (OSHA) promulgated the *Hazard Communication Standard* (*HCS*) (29 CFR 1910.1200) and applied it to manufacturing. Simply stated, the HCS required that employers:

- Develop a written Hazard Communications Standard (HCS) program,
- Maintain current lists of products containing hazardous chemicals on the worksite,
- Properly label containers of hazardous products,
- Maintain an archival file of Material Safety Data Sheets (MSDSs) for each of the hazardous products used on the worksite, and
- Provide training for workers in the proper procedures for storing, handling, and disposing of hazardous products.

When the HCS was promulgated, the breadth of the information management task was simply not anticipated and, as the Standard was implemented over the next decade, most manufacturers were overwhelmed to some degree by the scope of their HCS information management tasks. This was especially true with respect to the management of Material Safety Data Sheets (MSDSs) (Lindsay 1994). In spite of their efforts to comply with the HCS, most employers (and employees) did not feel a commensurate increase in access to the specific, accurate information actually needed (Hansen 1994).

In 1987 the Occupational Safety and Health Administration (OSHA) extended application of the Hazard Communications Standard (HCS) to cover new segments of the American workforce, including the construction industry. As a result, employers in construction were required to develop and implement hazard communication programs appropriate to their industry. In many respects implementation in construction was much more difficult than in manufacturing where the HCS was first applied. Some specific differences are:

- Large to medium sized construction businesses are organizationally decentralized. Administrative headquarters are seldom actually near a company's shifting, multiple worksites,
- The number of company employees (both managers and workers) at any particular worksite is generally small relative to a manufacturing operation,
- In construction a company's employees who work with hazardous chemicals regularly move from one worksite to the next,
- Construction companies in the US rely heavily on subcontractors whose employees are also very transitory with respect to particular worksites,

Perhaps, the most important difference distinguishing manufacturing from construction operations are their respective safety officers. Construction companies generally have an identified safety officer at their main office, but more often than not the safety officer assignment is secondary to other responsibilities. In most construction companies the person most likely to contend with the practical problems described above are on-site supervisors. Organizationally, a company's project superintendent is directly responsible for implementation of the company's HCS policies on container labeling, maintenance of required *Material Safety Data Sheet* (*MSDS*) files, employee training, maintenance of the site's list of hazardous chemicals, and keeping subcontractors informed.

In spite of the fact that construction industry employers, insurers, labor groups, and regulators generally agree that improved safety benefits everyone, the perceived dilemma is how to balance the daily trade-off between true costs of losses due to accident and health issues and the costs and inconveniences involved in delivering effective HCS safety information and training programs. Often from the (on-site) supervisor's perspective, the importance of the HCS program is not as important as making sure that subcontractors are moving through their tasks on schedule.

Aside from time management issues, field managers must deal with physical constraints. In practice, construction site field offices are cramped and busy. Although it is possible in a field office to keep files of MSDSs, maintain a list of on-site hazardous chemicals, to have blank, generic hazard warning labels, and to run small training programs, it is a very difficult task. The HCS files required on a commercial construction project can be very extensive and bulky. Establishing a regular filing system and keeping it current is time consuming. Providing training for employees every time a new chemical comes on-site is difficult, especially in a cramped field office. Under these circumstances it is easy for a superintendent to develop a perception of the HCS program as a regulatory nuisance and to put out only the minimum effort needed to meet OSHA requirements.

Workers can also perceive HCS programs as a nuisance, particularly if there are production-based incentive programs and they feel that HCS procedures waste their time. In addition, many workers perceive HCS programs negatively because of problems associated with limited literacy and English language skills (Adult Literacy in America, 1993). Safety materials are all too often written at levels and presented in ways that are not understandable to average workers (Rabin 1992). For example, studies indicate that average American workers cannot understand 40% of the information in a typical MSDS (Sattler, 1992). As a result, MSDS-based training can be frustrating or just plain boring to many workers.

In the construction industry, the result is that failures to comply with different aspects of the HCS top the list of OSHA compliance infractions. *The 100 Most Frequently Cited OSHA Construction Standards in 1991: A Guide for the Abatement of the Top 25 Associated Physical Hazards* (OSHA 1993) identifies the lack of a written HCS program, inadequate employee training, lack of MSDSs, lack of OHSA posters, and inaccessibility of MSDSs as the top five citations, respectively. This was still true as recently as 1997, during which OSHA continued 'to issue far more citations to employers for violating the Hazard Communications Standard than any other rule' (Karr 1997). It is noteworthy that the publication provides supplemental information on how to recognize and eliminate physical hazards, but offers no guidance on how to properly institute the HCS Standard. It is a difficult problem.

Although regular unannounced site visits by OSHA inspectors can insure that minimal compliance is a compelling issue on construction sites, inspections cannot address the simple organizational and logistical problems of HCS program maintenance on construction sites. The problem is how to streamline and simplify procedures from the superintendent's perspective. One must determine: how to maintain several simple databases, how to provide targeted, meaningful training, how to produce an appropriate hazard warning label with minimal hassle, and how to quickly find critical information in an emergency.

This paper reports on the development over the last four years of *HazCom Interactive*, an electronic HCS database project, that attempts to address the issues described above. In its first two years, this project's major goal was to develop procedures for porting information on hazardous chemicals from the IPCS International Chemical Safety Card (ICSC) system into a graphics-enhanced electronic HCS database to serve as a supplement to typical construction industry HCS programs. The database was designed specifically for delivering HCS information on typical construction-site PC computers. The project's specific initial objective was to develop a structured electronic database of multi-lingual ICSC information that included appropriately incorporated icons (or symbols), graphically indicating various types of hazards associated with chemicals and supporting audio.

In the last two years the project's emphasis shifted to training. After many demonstrations of *HazCom Interactive* to a wide range of audiences, it was apparent that even though the program was very favorably received, it did not readily fit into existing information delivery programs. In other words it was destined to become a passive resource that might occasionally be used. Apprenticeship educational program coordinators from the International Brotherhood of Painters and Allied Trades (IBPAT) recommended that the *HazCom Interactive* database be directly linked to interactive training programs covering appropriate topics, such as respirators and ventilation. In response to that suggestion, several interactive training modules were produced that cover different presentation styles. Each module links to the underlying database to provide detailed information about specific chemicals. This chapter reviews the underlying *HazCom Interactive* database and describes the interactive training modules developed for evaluation.

2 A PARTIAL SOLUTION

By conceptually recreating Hazard Communications Standard (HCS) programs as integrated electronic databases of information for use on field office PCs, many of the inconveniences and inefficiencies of the current system could be greatly reduced. In addition to compact storage, an electronic database makes locating and retrieving specific information easier. Information from an electronic HCS database could be accessed either on the computer screen or printed on demand. Multiple menus based on common names, active ingredients and scientific names would avoid questions about where to find information on a particular material. In addition to the menus, several standard search techniques (such as word wheels, hyperlinks, and full-text indexing) would be applied to the database to facilitate finding specific HCS information.

In addition to providing simple information access, an HCS database could readily generate labels and posters describing specific chemical hazards which could be printed on color stock and/or on color printers. Such labels and posters would use standard graphical symbols (Petreycik 1996) that indicate the type and degree of hazards associated with specific chemicals. Likewise, multi-media training programs derived from a HCS database could use the same graphical symbols on screen in conjunction with English (or second language) text and audio. This type of access to multi-media HCS programs would directly address some of the safety training problems discussed in a *Wall Street Journal* front page report (Labor Letter – December 21, 1993). Specifically, the article concluded that safety programming needs to incorporate more '*visually appealing materials*' and opportunities for '*participatory training*' to increase its effectiveness. All of the information delivery capabilities described above are consistent with the presence of PC computers in construction site field offices.

In practice most specific HCS program information currently used in the construction industry comes from Material Safety Database Sheets (MSDSs). Unfortunately, MSDSs are provided on paper (or as facsimile images) and are not generally available in digital form. Even worse, MSDSs are neither consistent nor uniform in their organization or appearance. The sheer volume of MSDSs further complicates the problem. The net result is that MSDSs are not a suitable source for building a HCS database.

At the opposite end of the spectrum, the International Chemical Safety Cards (ICSC) produced by the International Programme on Chemical Safety (IPCS 1991) are highly structured even to the extent that individual phrases used to characterize chemical properties and hazards are explicitly defined in English as well as a core group of other languages. The ICSC system is fundamentally different from the MSDS system mandated by the HCS standard. ICSCs cover individual hazardous chemicals while MSDSs cover individual products that contain hazardous chemicals. Manufacturers with minimal centralized coordination individually produce MSDSs while the ICSC system is highly coordinated by the IPCS.

3 HAZCOM INTERACTIVE: AN INFORMATIONAL DATABASE

Text Files: Text information from all currently available International Chemical Safety Cards (ICSC) contains information on: *Types of Hazards* (*Fire and Explosion*); *Types of Exposure* (*Inhalation, Skin, Eyes* and *Ingestion*); *Spillage Disposal*; *Storage, Packaging & Labelling*; and *Chemical Facts* (*Physical State, Chemical Dangers...*). In the ICSC database, the information is stored as a series of descriptive phrases that are explicitly linked to each information category. For example for the chemical Hydrazine under the category '*Physical State: Appearance*' the following descriptive phrases are listed; '*colourless*', '*fuming*', '*hygroscopic liquid*', '*with pungent odour*'. Text data (descriptive phrases) are available in many languages. The ICSC text data were imported into an object oriented database management system to support flexible access to the information and forms the foundation for *HazCom Interactive.*

Pictographs: The fields of information in the ICSCs were evaluated to prioritize data according to their need for graphical enhancement. For example pictographs for hazards such as fire and explosion and prevention measures such as '*NO open flames*' and '*ventilation*' were deemed essential. Conceptually, a *pictograph* (graphical symbol) database structure was developed to convey information that linked, as explicitly as possible, to IPCS glossary terms (phrases) as they were applied in the ICSCs to characterize chemicals. The pictographs adhere to international conventions intended to be understandable across cultures. To date, approximately 100 pictographs for the database have been developed in priority order.

Audio Files: In addition to the databases of text and graphical information, a database of audio files was created that linked explicitly to high priority text files. For example digital audio files for phrases such as '*Highly explosive*' and '*First rinse with plenty of water*' were recorded and stored.

User Interface: This project depends to a large extent on the quality of the user interface that displays the hazardous chemicals described. In particular the interface needs to present a visually appealing and participatory program that engages workers quickly and effectively delivers pertinent information (Szudy and Arroyo, 1994). The interface was designed to quickly and easily communicate ICSC information in a way that addresses the language and literacy requirements of its users (specifically through a combination of graphics, multi-language audio and multi-language text). It is a simple point and click interface with a variety of supplemental navigational features. Figure 1 shows the system's opening screen that defines major areas of information using text and pictographs. The figure cannot show the system's capability to activate audio files when the cursor is dragged across the various buttons. For example if the cursor was dragged across the uppermost pictograph, an audio file would be activated saying '*Chemical Search*'.

HazCom Interactive: Chemical Search (shown in Fig. 1) allows a user to search for a specific chemical by common name, scientific name or any of five chemical numbers (CAS,

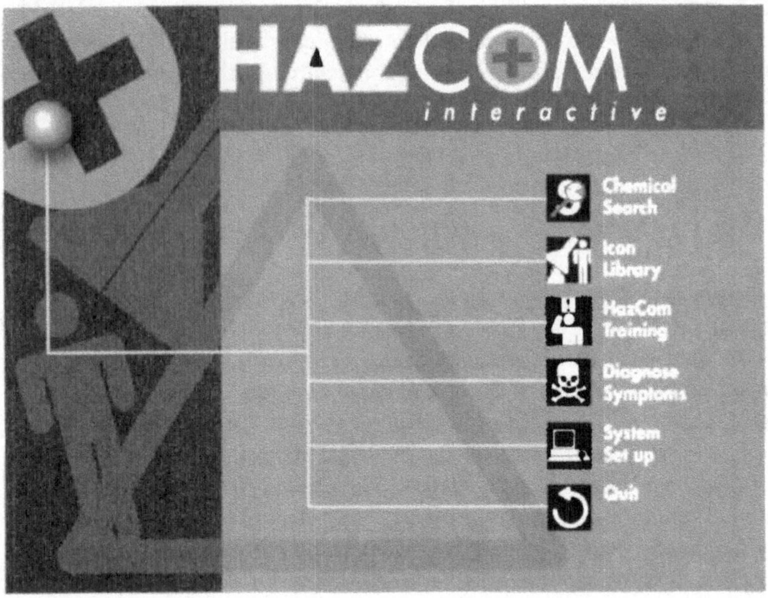

Figure 1. *HazCom Interactive* opening screen.

RTECS, ICSC, UN, or EC). Once a chemical is selected, such as *HYDRAZINE*, a general information screen allows the user to make specific information requests. Figure 2 shows the opening screen for *HYDRAZINE*. In addition to various chemical numbers, the opening menu offers access to seven categories of information: *Explosion Hazards, Emergency Medical, Fire Hazards, Exposure Hazards, Spillage Disposal, Storage, Packaging & Labeling, Chemical Facts & Additional Information*. Note that the seven category pictographs correspond to figures in the dark region on the screen's left that can be used as shortcuts to specific information.

To obtain more information on the effects of inhaling *HYDRAZINE*, the user selects *Exposure Hazards*; the program moves to a screen that displays the four exposure routes of *Inhalation, Skin, Eyes*, and *Ingestion*; *Inhalation* is chosen and a screen displays *Acute Symptoms, Prevention* and *First Aid*; and finally, *Acute Symptoms* is selected to display the descriptive phrases from the ICSC database that correspond to inhalation hazards associated with *HYDRAZINE*. Figure 3 shows the *Inhalation* screen with *Acute Symptoms* selected. The screen shows that *HYDRAZINE* can cause *Cough, Nausea, Burning sensation of nose and throat*, and/or *Convulsions*.

Multiple Languages: Because of the underlying design of the ICSC system, translation of the information between languages is greatly facilitated. Specifically, the phrases that correspond to the various categories are available in more than 18 languages including Dutch, French, German, Japanese, Spanish, French and others. Fully implemented *HazCom Interactive* could display text, deliver audio and print in all of these languages. Currently, it is fully functional in English and Spanish.

Other Features: In addition to displaying information on-screen, *HazCom Interactive* includes features to print information on demand. Specifically, graphical versions of Inter-

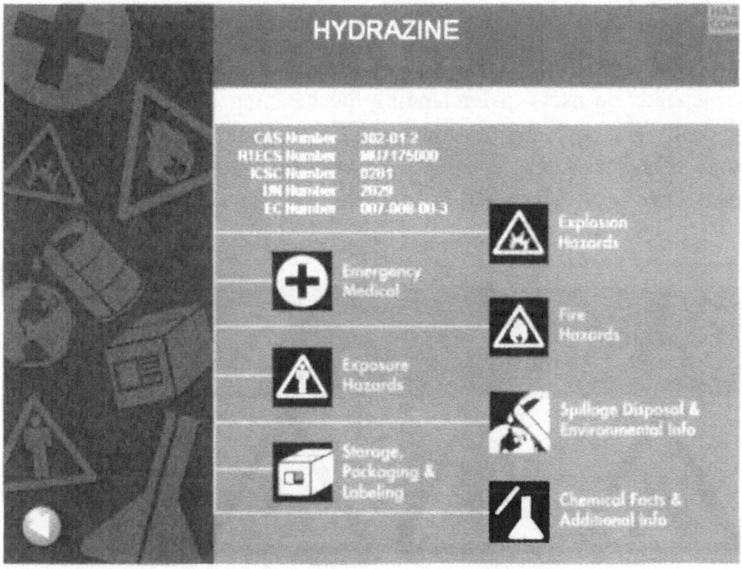

Figure 2. The opening screen for accessing information on HYDRAZINE.

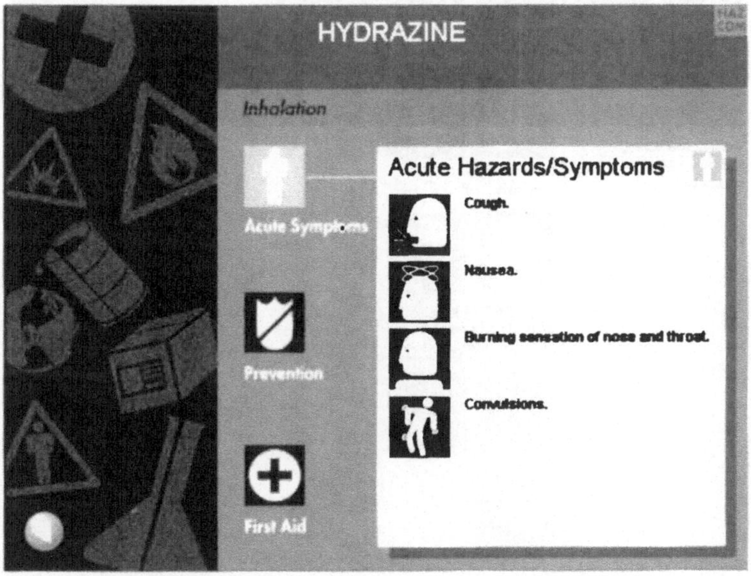

Figure 3. A screen showing the Acute Symptoms of HYDRAZINE inhalation.

national Chemical Safety Cards for specific chemicals can be printed. The graphical information card for Hydrazine is shown in Figure 4. In addition, container labels and posters can be printed. All of these print products rely heavily on the pictographs to convey information visually.

4 LINKAGES BETWEEN THE DATABASE AND TARGETED TRAINING MODULES

HazCom Interactive depends on users understanding the meaning of its graphical symbols. This can be partially accomplished through good, systematic design of the pictographs, however, some concepts cannot be simply represented symbolically. To support the pictographs, games have been developed to reinforce the meanings of symbols. The

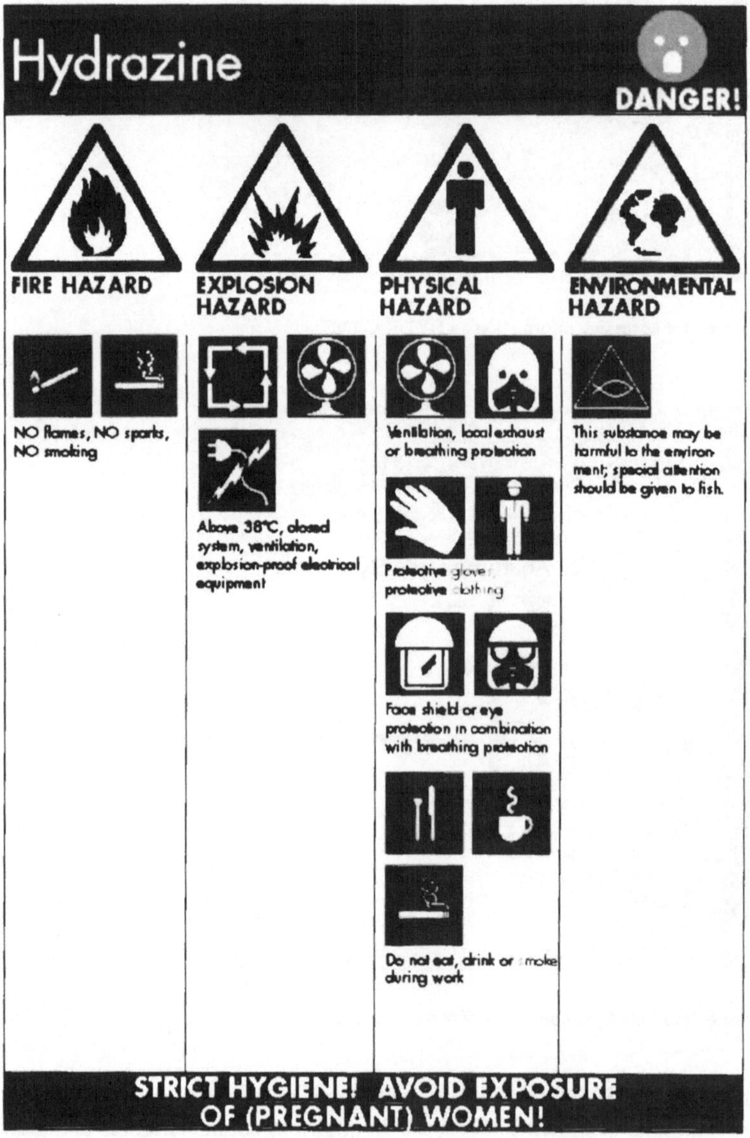

Figure 4. A print on demand information card for Hydrazine that heavily uses pictographs to convey hazards and preventative measures.

first game is intended to raise general awareness of the pictographs. Twelve random pictographs from the total collection are displayed in randomly selected slots. An example opening-screen from the game is shown in Figure 5. As the game starts, a phrase corresponding to one of the pictographs pops up in the window at the top of the screen, for example 'drowsiness'. As the words appear the user chooses what they think is the appropriate pictograph and clicks on it. If the selection is correct, the pictograph disappears and a new phrase pops up. A timer challenges the user to move as quickly as possible. If the user hasn't completed the game in 60 seconds, the game stops and the player is asked to play again. If the game is completed under 60 seconds, the player is congratulated and offered the opportunity to either play again to further improve his score or to exit the game. This is a simple, but highly interactive game and in presentations it is enthusiastically received.

The Matching Game shown in Figure 5 directly supports the *HazCom Interactive* database program by increasing users' recognition of the pictographs. It also supports use of pictographs on printed posters, cards and labels. However, it is very tightly linked to the *HazCom Interactive* database and isn't very suitable for use in more general training programs. To further increase use of the database, example supplemental training modules were produced that directly link to and from the database.

An example of how a primarily text based module can be linked to the *HazCom Interactive* database is shown in Figure 6. In this Ventilation training module several widely believed 'myths' are presented as true/false questions. After each question is answered an example that demonstrates the correct answer is displayed using text and graphics while a voice-over reads the text. In Figure 6 the text includes a reference to the fact that '*carbon monoxide*, found in the exhaust from gas-powered compressors has killed many painters.' In the program the text phrase *carbon monoxide* is hyper-linked into the *HazCom Interactive* data

Figure 5. An example opening-screen in the Matching Game used to train users in recognizing the pictographs used in the *HazCom Interactive* database.

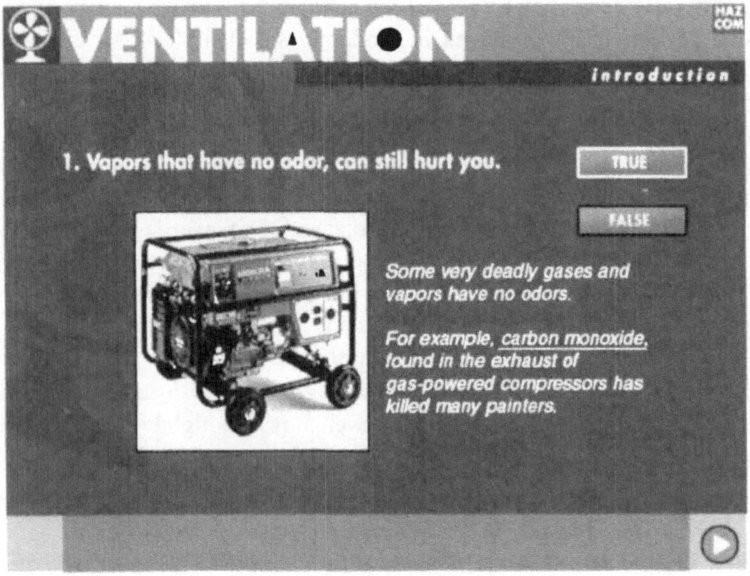

Figure 6. An example screen from a Ventilation training module that includes a reference to carbon monoxide that is hyper-linked to the *HazCom Interactive* database.

base where more detailed information on carbon monoxide (CO) is available. The user has the choice to access the additional information as needed. Ideally, training modules of this type could be developed to include many references and exercises directly linked to the database.

In addition to being able to link from training modules to *HazCom Interactive*, it is also possible to link from the *HazCom Interactive* database to a targeted training module. For example if hydrazine were going to be used on a site and its inhalation exposure hazards were being reviewed it could offer a direct link to training as shown in Figure 7. The training button could link to a menu of modules or straight to a specific module, for instance the carbon monoxide *HazCom Interactive* entry might link straight back to the true/false quiz described above.

The Prevention measures described for avoiding inhalation exposure to hydrazine (shown in Fig. 7), reference ventilation and breathing protection as two requirements. An example of a specific interactive training module that describes inlet and exhaust fan placement to maximize ventilation is shown in Figure 8. The module is presented as a simple game. Initially, both fans and the painter are placed outside the room. The user is asked to drag and drop the exhaust fan and then the supply fan into their proper locations by using the mouse. The dots are two colors, blue for good air and red for solvents. They move around in the room symbolically indicating the fans' effect on local solvent concentrations. Once the fans are positioned, the user is asked to place the 'painter' in the correct area to work. If the user tries to position the worker in the wrong location, the 'painter' won't drop.

The fan placement module shown in Figure 8 is concerned with ventilation. The inhalation exposure prevention advice for hydrazine also suggested that breathing protection was needed. Another training module covers personal protective equipment (PPE) in-

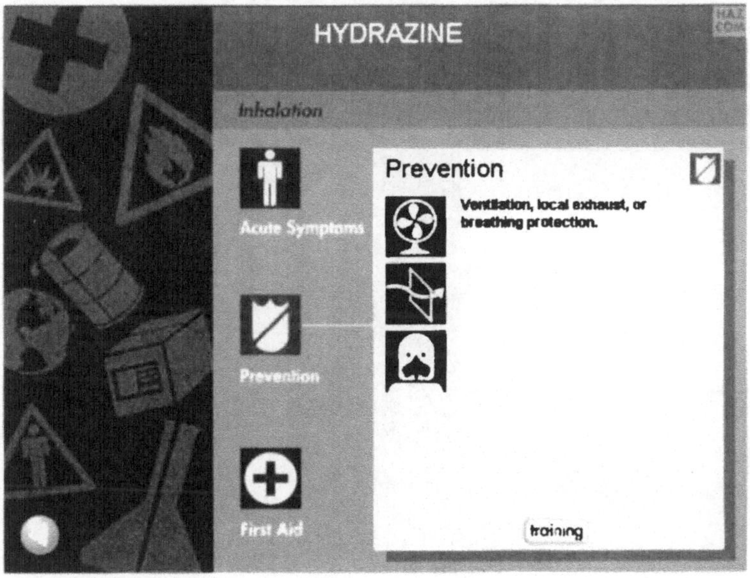

Figure 7. A *HazCom Interactive* inhalation exposure prevention screen with training button that links to specific supplemental training modules.

cluding breathing protection devices. This module presents the user with a work scenario, such as 'You will be spray painting an indoor room with no windows and only one door with latex paint.' The user is asked to dress and properly equip himself for doing the job. The computer screen shows a supply locker with various PPE categories. The user selects a category and is offered options. In Figure 9 the user has already chosen hearing protection, clothing, shoes and gloves and is preparing to choose from various respiratory protection devices.

Once the user has completed his selections, he can try to go to work by clicking on the arrow to the left. If the user has selected at least minimally acceptable protection then he is congratulated and the game concludes. If the user has made any unacceptable choices then he is offered an explanation of why particular selections are inadequate and sent back to the locker room to fix the problems. Obviously, the worker shown in Figure 9 will need something sturdier than flip-flops.

5 CONCLUSIONS

HazCom Interactive is based on a highly modularized series of printed cards – the International Chemicals Safety Cards produced by the International Programme on Chemical Safety. In fact, as a print product it looks very much like a 'database' highly suitable for a traditional computer application. Of course, the project described in this chapter is not a traditional database application, rather it's concerned with modifying an existing database to improve *actual* access and to expand its use by directly linking it to targeted training modules.

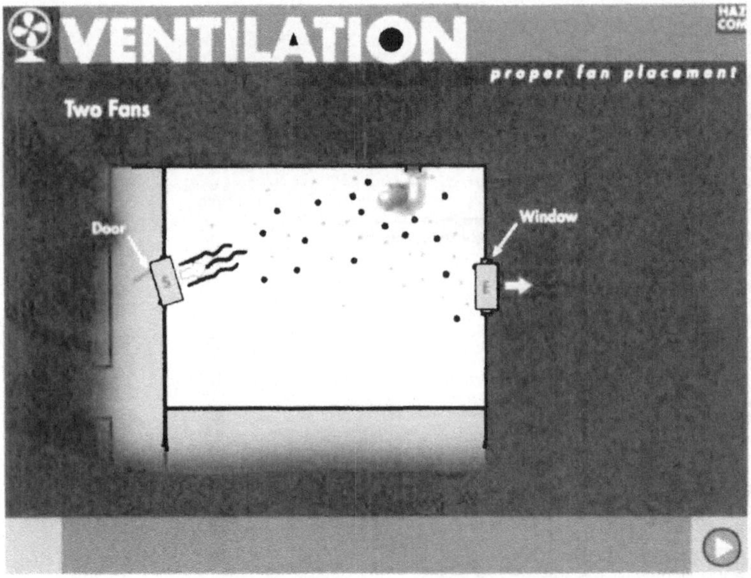

Figure 8. A specific training module on properly placing supply and exhaust fans to minimize solvent concentration in a worker's breathing zone.

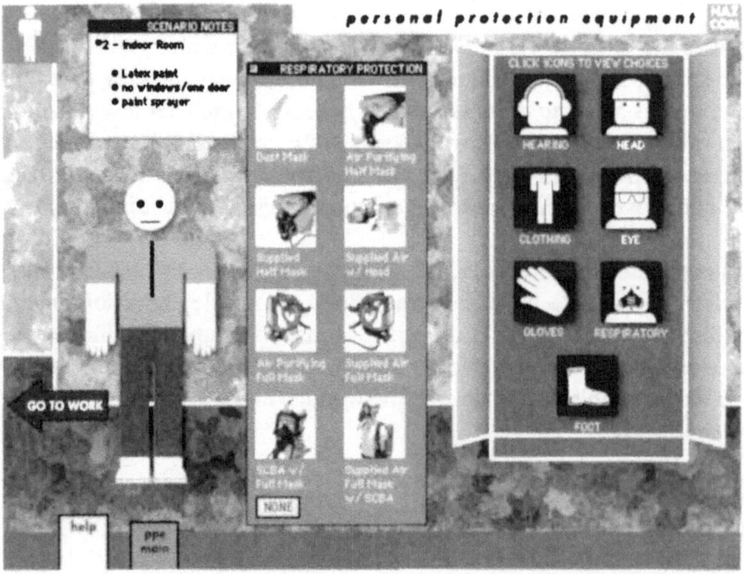

Figure 9. A screen from the PPE game showing the choices of respiratory protection that the user must select from to be suitably equipped for a particular job.

The system's navigational schemes are intended to allow users to find specific information quickly (and in a nonlinear fashion). The use of graphics and audio files is targeted to expanding access to semi-literate users. The print-on-demand capabilities are intended to expand the impact into the physical workplace. The system also creates a broad foundation on which training programs can be developed in an environment where a critical mass of information can be found.

HazCom Interactive has all of the advantages of other electronic database systems. Nothing is ever misfiled. The same information can be accessed from many different directions (menus). Version control can be insured. It takes up very little space and is cheap to replicate and distribute. The primary disadvantage is that users need computer hardware to access the information. As computers proliferate that's unlikely to be a serious constraint. The project is expected to continue for one more year during which time its usefulness and practicality are expected to steadily increase until it is delivered to and in use on construction worksites.

Even though *HazCom Interactive* represents a conceptual improvement over current HCS program implementations, it still has a major flaw: *HazCom Interactive* is based on chemicals not products. Furthermore, the underlying database is not exhaustive and even if the database contained all hazardous chemicals that would not address the problem of mixtures. OSHA requires that manufacturers be accountable for their products and be responsible for producing MSDSs on a product by product basis. Under the current regulations, even products that contain a single hazardous chemical could not be covered by *HazCom Interactive*.

REFERENCES

Adult Literacy in America: National Adult Literacy Survey. Office of Educational Research and Improvement. Washington, DC: US Department of Education, September 1993.

Hansen, M.D. 1994. 'Environmental Safety and Health Compliance in the 1990s: Using Technology to Manage the Paper Chase.' *Professional Safety*. American Society of Safety Engineers, 1800 East Oakton St. Des Plaines, IL. 39(2): 29-32.

International Programme for Chemical Safety. 1991. 'International Chemical Safety Cards.' Commission of the European Communities, Telecommunications, Information Industries and Innovation. Luxembourg.

Karr, Al. 1997. 'Top 10 OSHA Violations: Same Old Tune?' *Safety + Health* 156(6).

Lindsay, Ron. 1994. 'Getting a Handle on MSDS Records'. *Industrial Hygiene News* 17(4).

Naisbett, J. 1982. Megatrends. Warner. New York.

Occupational Safety and Health Administration. 1993. 'The 100 Most Frequently Cited OSHA Construction Standards in 1991: A Guide for the Abatement of the Top 25 Associated Physical Hazards.' US Department of Labor, Washington, DC.

Petreycik, R. M. 1996. 'EU Cracks Down on Chemical Labeling'. *Safety + Health*. March pp. 68-71.

Rabin, A.T. 1992. 'Consequences of Adult Illiteracy Reach Far into the Marketplace.' *Occupational Health and Safety*, June 1992 pp. 36-42.

Sattler, B. 1992. 'Rights and Realities: A Critical Review of the Accessibility of Information on Hazardous Chemicals.' *Occupational Medicine: State of the Art Reviews*, April-June 1992, pp. 189-96.

Szudy, E. & M.G. Arroyo. 1994. 'The Right to Understand: Linking Literacy to Health and Safety Training.' Labor Occupational Health Program, University of California at Berkley.

Biographical sketches of scholars

JOHN ALEXANDER (USA)

John Alexander is a Professor in the Department of Urban and Regional Planning and director of the GeoPlan Centre at the University of Florida. His major research interests include Geographical Information Systems (GIS), and dual imaging. Through his development of hand held dual imaging systems high risk areas like trenches and high power lines can be avoided as far as measurement thus keeping employees away from such hazards. He has done considerable research for IBM, Trimble and Motorola in the area of electronics. His Ph. D. focussed primarily with environment issues in an electronic format which is consistent with his concern for construction safety, health and the environment.

MICHAEL BAXLEY (USA)

Michael Baxley is a medical doctor who is the medical director at Prudential Insurance in Jacksonville, Florida. He has conducted considerable research for the Navy Seals. As part of his training in becoming an occupational doctor his studies also have involved injuries and methodologies that are used to reduce fraudulent injury claims from a medical point of view. He has worked closely with the insurance industry in developing case systems to identify fraudulent injuries and has developed a database in this regard.

JOACHIM BERGER (GERMANY)

Joachim Berger is a leading manager with Bau-BG in Frankfurt, Germany and has extensive experience and interest in safety and health issues in both Germany and the European Union (EU). In particular he is interested in the organisation of occupational safety and health in Germany, preventative health protection strategies, sick building syndrome (SBS), safety planning in the design, construction and utilisation phases of projects and international co-operation in respect of labour and workers' compensation.

RICHARD J. COBLE (USA)

Richard J. Coble is an Associate Professor in the ME Rinker Sr School of Building Construction and the Director of the Centre for Construction Safety and Loss Control at the University of Florida. He has extensive hands-on experience in construction having undertaken several major construction projects throughout the USA. His major research interest is in safety and health in construction and has recently been conducting investigative studies into workman's compensation fraud. He has shown a strong research interest in the area of automating the construction foreman which is integral to scheduling safety into all aspects of the construction process. He is currently the international director of CIB W99 which is an international consortium of construction safety experts. He has published widely in the area of safety and health.

LUÍS ALVES DIAS (PORTUGAL)

Luis Alves Dias is a Professor in the Civil Engineering Department at Institute Superior Technico in Lisbon, Portugal. His current teaching at the Institute is specifically directed to the areas of construction safety and health. He is the head of several safety organisations in Portugal and has the premier safety and health consultancy in the country. He is an advisor to the Portuguese government on safety and health issues. He has authored several books and journal articles specifically related to construction safety, health and the environment.

ROY DUFF (UNITED KINGDOM)

Roy Duff is one of the premier construction researchers at the University of Manchester Institute of Science and Technology in the United Kingdom. His research has included investigating the impact of behaviour and its modification on safety and quality performance in construction. He has done considerable workforce studies related to worker safety and health which has produced numerous international peer reviewed articles.

THEO HAUPT (SOUTH AFRICA)

Thee Haupt is a lecturer at the Department of Construction Management and Quantity Surveying at Peninsula Technikon, Cape Town, South Africa. He has served as the Chairperson of the Western Cape branch of the South African Institute of Building (SAIB). He remains a National Council member of SAIB and enjoys membership of Architects and Surveyors Institute, Chartered Institute of Building, Commonwealth Association of Surveying and Land Economics. His research interests include infrastructure policy and delivery in the context of developing countries. However, he is presently engaged in doctoral studies at the University of Florida where his focus has been on construction safety issues. He has published several safety related articles and conference papers.

ROBERT F. HERRICK (USA)

Bob Herrick is a lecturer on industrial hygiene at the Harvard School of Public Health and has research and teaching interests in exposure-reactive aerosols, characterisation of complex exposures, and the interaction of individuals with exposure sources. He has more than 20 years expertise working directly for and with NIOSH. He has published widely in areas of health related issues in the construction industry.

PIERCE JONES (USA)

Pierce Jones is a Professor in Agricultural Engineering at the University of Florida. He is involved in consultancy business which specialises in safety and health training and implementation in the construction industry. He has developed a CD-ROM database for National Institute for Occupational Safety and Health (NIOSH) and has been involved extensively with CPWR. His involvement with continued education for I.S.I.S at the University of Florida has afforded him the opportunity to provide training for many of the farm bureau outlets throughout the state. The background he had previously conducted in management information systems has allowed him to develop a database specialisation in regards to safety and health.

XILA LIU (CHINA)

Xila Uu is the Professor and Head of the Department of Civil Engineering at Tsinghua University. He is also a professor at several other universities in China. His many awards include the Research Award of the National Science Congress of China as well as the ASCE Raymond C. Reese Research Prize. He has authored or co-authored more than 200 articles and is a contributing editor of both the Journals of 'Engineering Mechanics' and 'Fuzzy Systems and Mathematics'. His research interests include constitutive modelling of concrete, safety analysis of structures, durability modelling of structures, knowledge-based systems for structural design and damage assessment. Currently he is tasked with reducing construction worker safety accidents and fatalities.

RON SHARPE (AUSTRALIA)

Ron Sharpe is the National Manager, Construction Process Systems at CSIRO Building Construction and Engineering, Melbourne, Australia which he joined in 1969. Recent highlights of his research management has focused in the areas of public safety and health systems which include the development of Building Code of Australia Expert System (BCAider), construction process re-engineering, development of pattern recognition software for robots using expert systems and neutral networks, establishment of Australian component of the Centre for Sustainable Urban and Regional Development (CSURD) in Indonesia and setting up the Australasian Chapter of International Alliance for Interoperability (IAI) developing a universal information language enabling all elements of the construction industry to seamlessly communicate across all segments and around the globe.

PAM SUSI (USA)

Pam Susi is an industrial hygienist with considerable experience in the construction industry serving as the Exposure Assessment Program Director for the Center to Protect Workers' Rights (CPWR) since 1993. She has been the project co-ordinator for CPWR's 'Task-based Exposure Assessment Model (T-BEAM) for Construction'. She authored a chapter on non-residential structures in the development of 'Worker Protection Practices for Lead Abatement. She was responsible for developing 'Model Specifications for the Protection of Workers from Lead on Steel Structures'. She currently manages a NIOSH funded project aimed at implementing and evaluating these specifications.